Eclipse Almanac
2061 To 2070

Ten Years of Solar and Lunar Eclipses

Color Edition

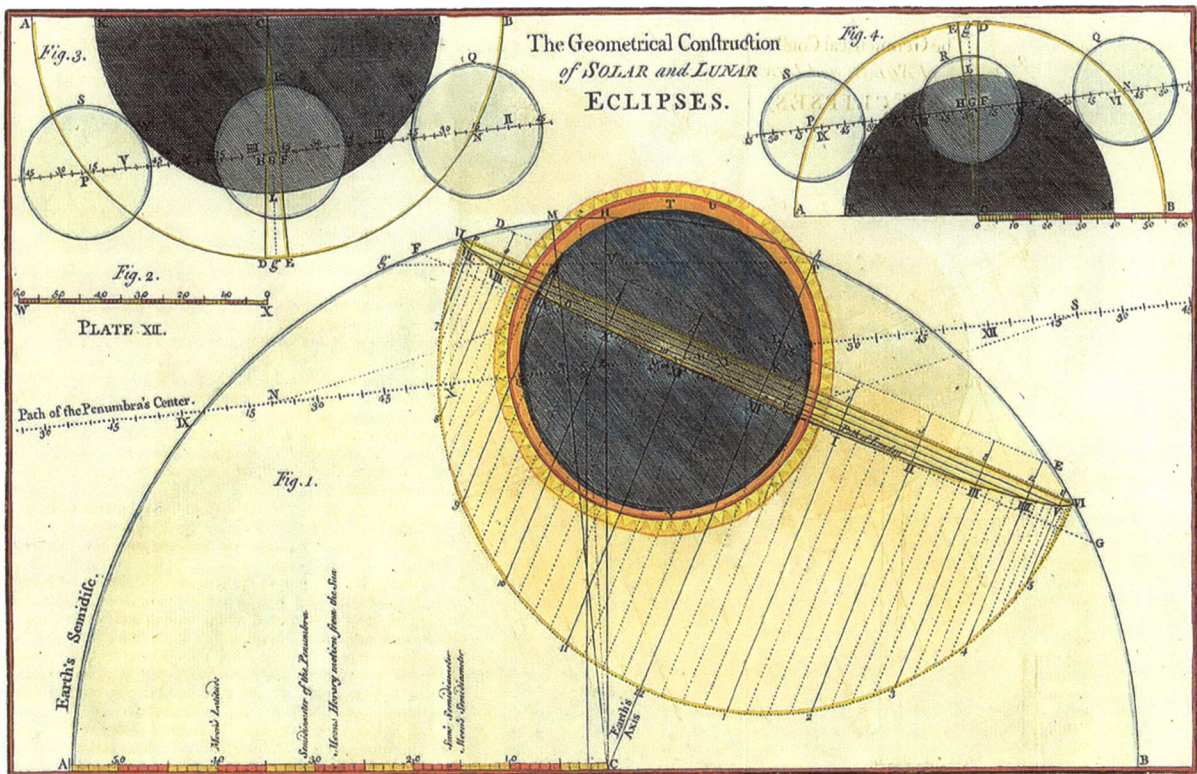

The Geometrical Construction of SOLAR and LUNAR ECLIPSES.

Fred Espenak

Edition 1.0
October 2020

Eclipse Almanac 2061 to 2070 – Color Edition

Ten Years of Solar and Lunar Eclipses

Astropixels Publishing
P.O. Box 16197
Portal, AZ 85632

www.astropixels.com/pubs

This book may be ordered at: *www.astropixels.com/pubs/EclipseAlmanac.html*

More about solar and lunar eclipses from 2061 to 2070 can be found at EclipseWise:

www.astropixels.com/news/EclipseAlmanac.html

Astropixels Publication Number: AP034

First Edition (Version 1.0a)

ISBN 978-1-941983-34-8

Printed in the United States of America

Front Cover: "The Geometrical Construction of Solar and Lunar Eclipses," James Ferguson (1756).

Back Cover: Portrait of Fred Espenak (Copyright ©2018 by Fred Espenak).

Preface

The *Eclipse Almanac* contains maps and diagrams of every solar and lunar eclipse over a ten-year period. This permits the reader to look ahead and easily determine when and where each of these events will be seen. Particular details about each eclipse are included as well as a 25-year table looking further into the future.

Section 1 covers solar eclipses, while Section 2 is devoted to lunar eclipses. Brief explanations are given for the different types of eclipses, and descriptions of the visual appearance of each one is included. Section 3 tabulates the date and time of the Moon's phases over a decade. New Moon and Full Moon phases that coincide with solar and lunar eclipses, respectively, are identified.

The *Eclipse Almanac* series consists of five volumes. Each one covering a single decade as follows:

1) Eclipse Almanac 2021 to 2030 4) Eclipse Almanac 2051 to 2060
2) Eclipse Almanac 2031 to 2040 5) Eclipse Almanac 2061 to 2070
3) Eclipse Almanac 2041 to 2050

Each volume is available in a color edition as well as a more economical black and white edition.

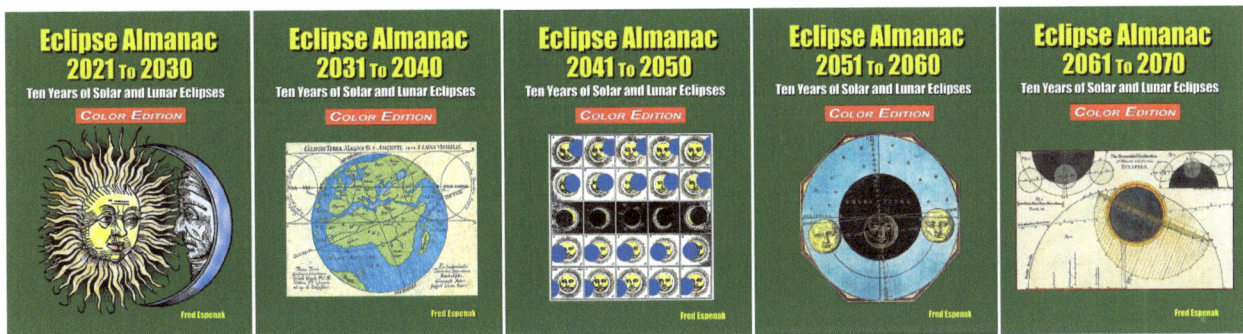

All times listed in the *Eclipse Almanac* are in Universal Time (UT1). This is the modern replacement of Greenwich Mean Time, which was based on mean solar time from Greenwich, England. In comparison, Universal Time is based on Earth's rotation with respect to distant quasars.

For North Americans, the conversion from UT1 to local time is as follows:

Atlantic Standard Time (AST) = UT1 - 4 hours
Eastern Standard Time (EST) = UT1 - 5 hours
Central Standard Time (CST) = UT1 - 6 hours
Mountain Standard Time (MST) = UT1 - 7 hours
Pacific Standard Time (PST) = UT1 - 8 hours

If Daylight Saving Time is in effect in the time zone, you must ADD one hour to the standard time.

The conversion of Universal Time to other time zones is easily found on the Internet.

For more information about solar and lunar eclipses from 2021 to 2070, visit *EclipseWise.com* at:

www.astropixels.com/news/EclipseAlmanac.html

Table of Contents

Central Solar Eclipses from 2061 to 2070

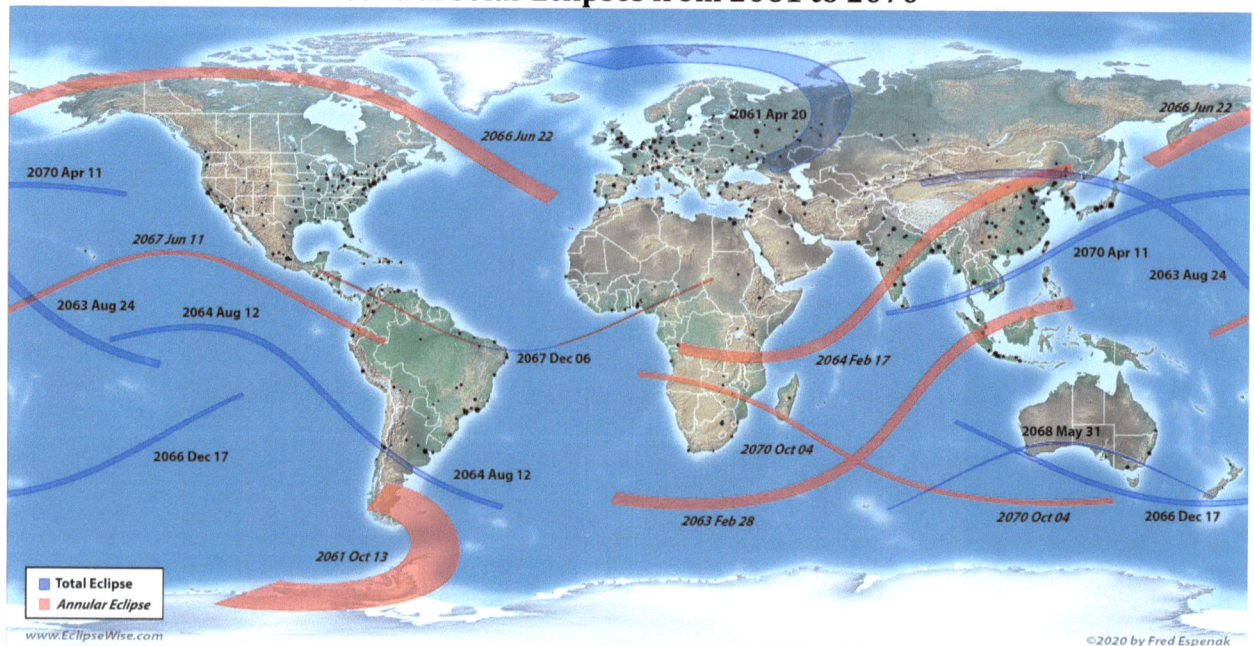

The path of every central solar eclipse from 2061 to 2070 is plotted on a world map. The central paths of total eclipses are shaded blue, while annular eclipses are shaded red. For hybrid eclipses, part of the path is shaded blue (total), and part is shaded red (annular). Major cities are plotted as black dots, scaled by population size. (©2020 F. Espenak)

Photo 1–1 The "Diamond Ring Effect" is seen just before totality begins. Total Solar Eclipse of 2018 Jul 03. ©2018 F. Espenak

Section 1: Solar Eclipses

Introduction

The Moon orbits Earth once every 29.5306 days with respect to the Sun. Over the course of its orbit, the Moon's changing position relative to the Sun results in its familiar phases: New Moon > First Quarter > Full Moon > Last Quarter > New Moon. The New Moon phase is the only one not visible because the illuminated side of the Moon then points away from Earth.

The orbit of the Moon is tilted about 5.1° to Earth's orbit around the Sun. The points where the two orbits appear to cross are called the nodes. When the New Moon occurs near one of these nodes, the Moon's shadow falls on a portion of Earth and a solar eclipse is visible from that region.

The Moon's shadow is composed of three cone-shaped components. The outer or penumbral shadow is a zone where the Sun's rays are partially blocked. Nested within the penumbra is the umbral shadow — a region where direct rays from the Sun are completely blocked. The conical umbra tapers to a point beyond which extends an expanding cone called the antumbra. From within this third shadow, the Moon appears smaller than the Sun and is seen in silhouette against the solar disk.

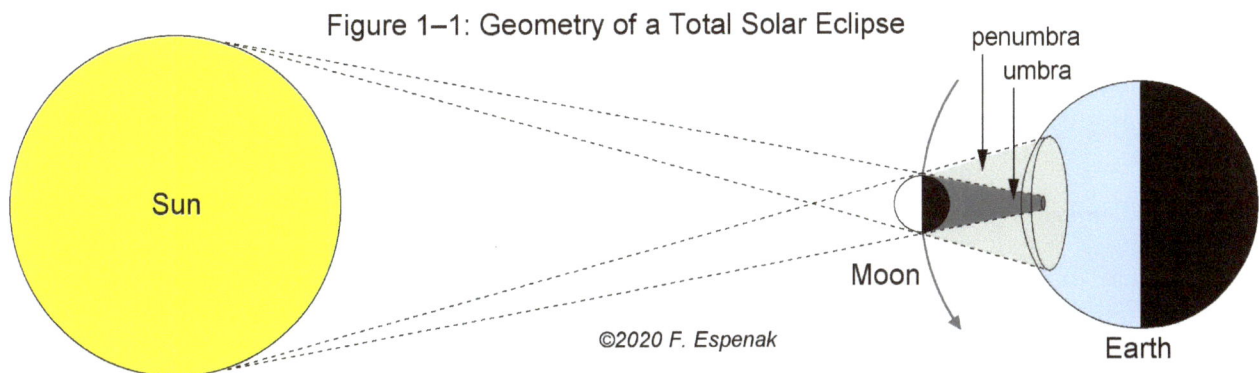

Figure 1–1 illustrates the geometry of a total solar eclipse. A partial eclipse is visible from within the large penumbral shadow, while the total eclipse is only seen from the much smaller umbral shadow.

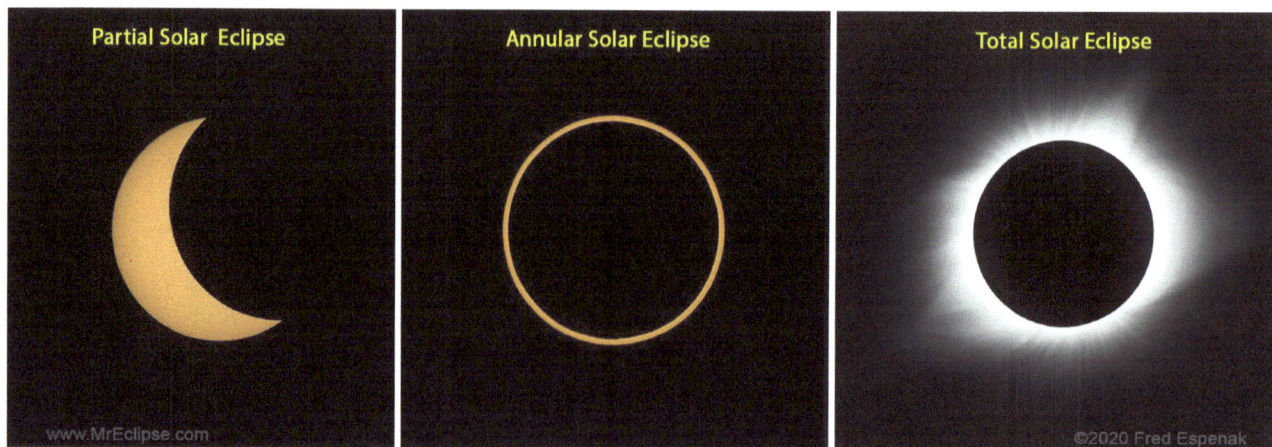

Photo 1–2 Examples of a partial, annular, and total solar eclipse. ©2020 F. Espenak

Types of Solar Eclipses

There are four types of solar eclipses:

1. **Total Solar Eclipse** — The Moon's penumbral and umbral shadows traverse Earth. The Moon's antumbral shadow extends beyond Earth's surface. The Moon appears larger than the Sun and completely covers the solar disk when viewed from within the umbral shadow. A partial eclipse is seen within the penumbral shadow (Figure 1–1).
2. **Annular Solar Eclipse** — The Moon's penumbral and antumbral shadows traverse Earth. The Moon's umbral shadow completely misses Earth. The Moon's disk appears smaller than the Sun so a bright ring of the Sun's disk surrounds the Moon when viewed from within the antumbral shadow. A partial eclipse is seen within the penumbral shadow (Figure 1–2).
4. **Hybrid Solar Eclipse** — The Moon's penumbral, umbral and antumbral shadows all traverse parts of Earth. The curvature of Earth's surface brings some regions into the umbra and others into the antumbra. The eclipse is total within the umbra and annular within the antumbra. A partial eclipse is seen within the penumbral shadow. Hybrid eclipses are also known as annular-total eclipses.
4. **Partial Solar Eclipse** — The Moon's penumbral shadow traverses Earth while the umbral and antumbral shadows miss Earth. A portion of the Sun's disk is obscured from within the penumbra.

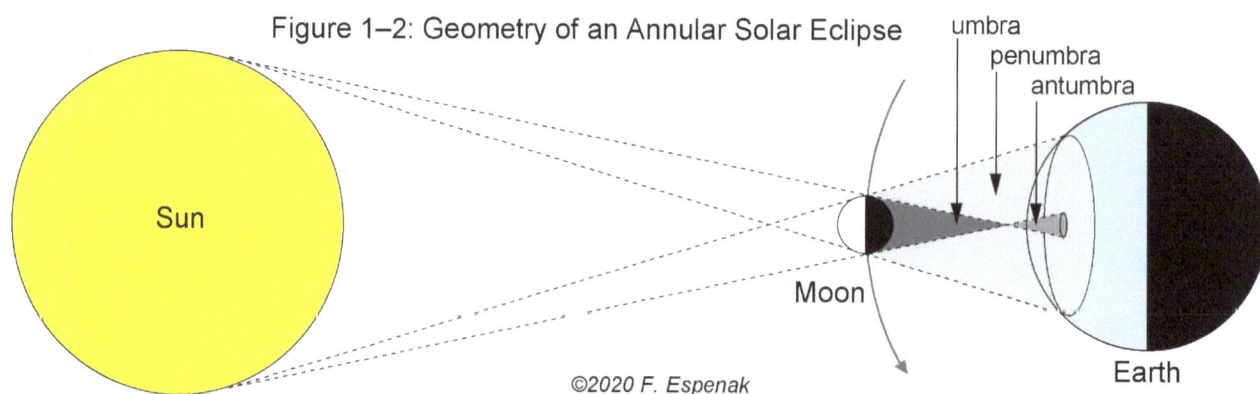

Figure 1–2 illustrates the geometry of an annular solar eclipse. A partial eclipse is visible from within the large penumbral shadow, while the annular eclipse is confined to the much smaller antumbral shadow.

Annular, total and hybrid eclipses are sometimes referred to as *central* eclipses[1]. However, on rare occasions it is possible to have an annular or total eclipse that is non-central. This occurs in Earth's polar regions when only the edge of the umbral or antumbral falls on Earth while the shadow axis misses the planet entirely. The eclipse of 2043 April 09 is a non-central total solar eclipse.

---

[1] A central eclipse is one in which the central axis of the Moon's umbral/antumbral shadow intersects with Earth's surface.

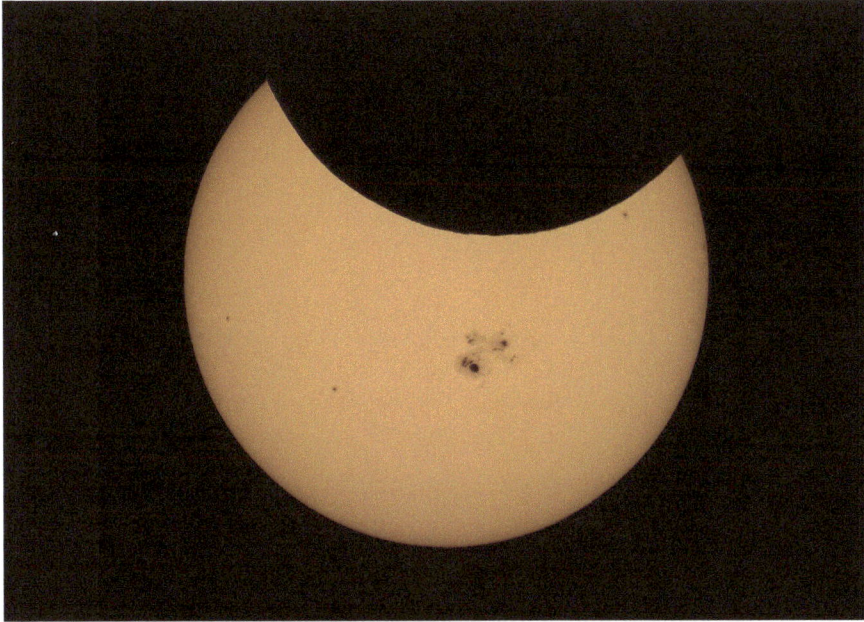

Photo 1–3 Partial solar eclipse of 2014 Oct 23. ©2014 F. Espenak

Visual Appearance of Partial Solar Eclipses

When the Moon's penumbral shadow strikes Earth, a partial eclipse of the Sun is visible from that region. The apparent motion of the Moon with respect to the Sun is gradual — the partial phases can two hours or more. During this time, the Moon's dark limb slowly creeps across the Sun's disk.

Partial eclipses cannot be viewed with the unprotected eye because the Sun is still extremely bright. Special techniques are needed to safely view the eclipse (for example, special eclipse filters or glasses). Even when a partial eclipse reaches its maximum phase, the sky and landscape remain bright. But careful inspection of dappled sunlight beneath a leafy tree will reveal multiple images of the eclipse. The gaps between the tree leaves act like pinhole cameras and project images of the crescent Sun onto the ground.

The Moon's penumbral shadow is typically 4200 to 4500 miles (6700 to 7300 kilometers) in diameter and can cover a significant fraction of the daytime hemisphere of Earth. Consequently, partial eclipses are visible from large geographic regions as the penumbra sweeps across Earth's surface.

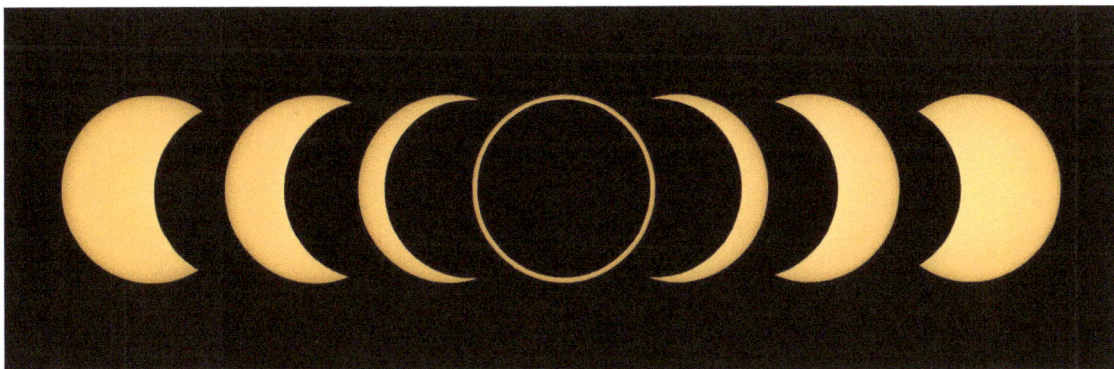

Photo 1–4 shows various phases of the Annular solar eclipse of 2005 Oct 03. ©2005 F. Espenak

Visual Appearance of Annular Solar Eclipses

During an annular eclipse, the Moon's penumbral and antumbral shadows sweep across Earth. Compared to the penumbra, the antumbra is much smaller and has a maximum diameter of 232 miles (374 kilometers). Because of this, the antumbra covers a tiny fraction of Earth's surface.

A partial eclipse is visible within the penumbral shadow, but only observers located in the much narrower track of the antumbra will see an annular eclipse. For this reason, the antumbra's trajectory across Earth is called the path of annularity.

All annular eclipses begin with a series of partial phases lasting an hour or more. At the peak of the eclipse, the Moon's disk can be seen in complete silhouette against the Sun. The remaining solar photosphere appears as an intensely bright ring of light surrounding the Moon. The annular phase can last a maximum of 12 ½ minutes but is more typically 3 to 5 minutes in length. After annularity, another series of partial phases occur as the Moon gradually uncovers the Sun.

Special precautions must be used to watch an annular eclipse (just like partial eclipses). Even during the annular phase, the Sun is dangerously bright and cannot be viewed without a solar filter. The landscape and sky remain bright throughout the eclipse, giving little indication of the celestial event in progress.

Photo 1–5 The Sun's corona is only visible to the naked eye for a few minutes during a total eclipse of the Sun. This is an HDR composite image of the total solar eclipse of August 21, 2017 from Casper, Wyoming.
© 2017 F. Espenak, www.MrEclipse.com

Visual Appearance of Total Solar Eclipses

During a total eclipse, the Moon's penumbral and umbral shadows fall upon Earth. The umbra has a maximum diameter of 170 miles (273 kilometers). The narrow track traced out as the umbra sweeps across Earth's surface is called the path of totality. The Sun is completely obscured by the Moon from within this zone. The total phase can last up to 7 ½ minutes but is more typically 2 to 3 minutes in length.

Total eclipses all begin and end with a series of partial phases lasting an hour or more. But this is where the resemblance to partial and annular eclipses ends — the total phase is the most spectacular astronomical event visible to the naked eye. During totality, the Sun's outer atmosphere — the solar corona — appears as a gossamer halo surrounding the Moon, and bright stars and planets are visible.

The eclipse takes on a unique character about five minutes before the total phase commences. The Sun has a foreboding quality and casts abnormally sharp shadows. The approaching lunar umbra darkens the western sky and the air temperature is noticeably cooler. A minute before totality, pale shadow bands[2] ripple across the ground.

The ambient light grows feeble even though the crescent Sun is still too bright to see. In the final seconds, the Sun's corona emerges from the glare as the solar crescent shrinks to a brilliant jewel. This celestial diamond ring lingers for a moment before the sunlight is extinguished and totality begins.

The Sun's glorious corona is now displayed to full advantage in the darkened sky. Standing within the Moon's umbra affords the rare and unprecedented opportunity to gaze directly at the glowing million-degree plasma surrounding our star. Twisted, tortured, and constrained by the Sun's enormous magnetic fields, the solar corona is revealed to the naked eye only during the brief seconds when the Moon completely blocks the brilliant disk of the Sun. An eerie twilight bathes the landscape and the colors of dusk surround the horizon.

Minutes race by like seconds. Suddenly, a sparkling bead of sunlight reappears along one edge of the Moon and quickly grows to blindingly bright proportions. Daylight returns as the corona fades and the totality ends. Another hour of partial phases remains before the eclipse is over.

While filters are required for viewing the partial phases, they must be removed for totality. The total phase is the only time it is completely safe to look directly at the Sun without protection. In fact, the total phase is not even visible through solar filters because the Sun's corona is a million times fainter than the photosphere[3].

Figure 1–3: Solar Eclipse Contacts

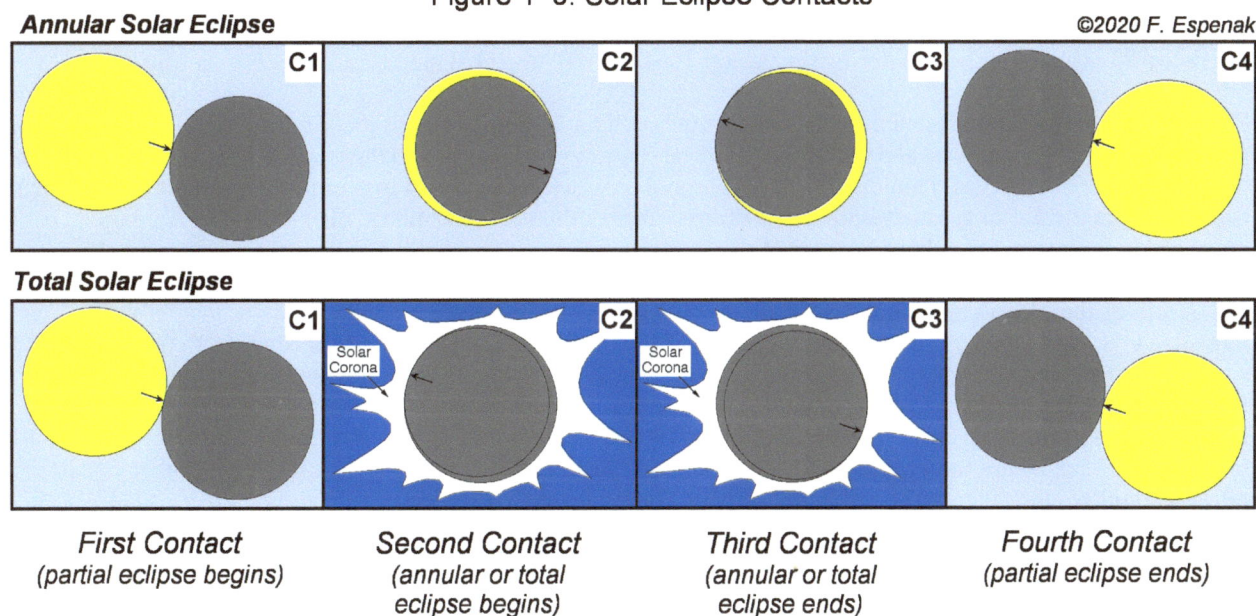

Figure 1–3 illustrates the four contacts for annular and total solar eclipses. The arrows indicate the contact point of the Moon with the Sun's limb in each diagram.

Solar Eclipse Contacts

During the course of a solar eclipse, the instants when the edge of the Moon's disk becomes tangent to the Sun's disk are known as eclipse contacts. They mark various stages or phases of a solar eclipse (Figure 1–3).

Partial solar eclipses have two primary contacts.

First Contact (C1) — Partial Eclipse Begins (Instant of first exterior tangency of the Moon with the Sun)
Fourth Contact (C4) — Partial Eclipse Ends (Instant of last exterior tangency of the Moon with the Sun)

---

[2] Shadow bands are caused by the thin solar crescent illuminating Earth's atmosphere before and after totality.
[3] The photosphere is the visible surface of the Sun's disk.

Central solar eclipses (total, annular or hybrid) have four primary contacts. Contacts C2 and C3 mark the instants when the Moon's disk is first and last internally tangent to the Sun. These are the times when the annular or total phase of the eclipse begins and ends, respectively.

First Contact (C1) — Partial Eclipse Begins (Instant of first exterior tangency of the Moon with the Sun)
Second Contact (C2) — Total (or Annular) Eclipse Begins (Instant of first interior tangency of Moon with Sun)
Third Contact (C3) — Total (or Annular) Eclipse Ends (Instant of last interior tangency of Moon with Sun)
Fourth Contact (C4) — Partial Eclipse Ends (Instant of last exterior tangency of the Moon with the Sun)

Photo 1–6 Sequence of partial phases, Diamond Rings, and Totality. Total Solar Eclipse of 2017 Aug 21. ©2017 F. Espenak

Explanation of Global Solar Eclipse Maps

There are 23 eclipses of the Sun during the period 2061 to 2070. A global map is included for each eclipse.

The geographic visibility of each eclipse is illustrated with an orthographic projection map of Earth showing the path of the Moon's penumbral and umbral/antumbral shadows with respect to the continental coastlines, political boundaries (circa 2016), and major cities (represented by black dots). North is up and the daylight terminator is drawn for the instant of greatest eclipse. An asterisk marks the sub-solar point where the Sun appears directly overhead at that time. The salient features of the eclipse maps are identified in the key diagram on page 11.

The limits of the Moon's penumbral shadow delineate the region where a partial solar eclipse is visible. This irregular or saddle shaped region often covers more than half the daylight hemisphere of Earth and consists of several distinct zones or limits. At the northern and/or southern boundaries lie the limits of the penumbra's path. Partial eclipses have only one of these limits, as do central eclipses when the Moon's shadow axis falls no closer than about 0.45 radii from Earth's center. Great loops at the western and eastern extremes of the penumbra's path identify the areas where the eclipse begins/ends at sunrise and sunset, respectively. If the penumbra has both a northern and southern limit, the rising and setting curves form two separate, closed loops (e.g., 2024 Apr 08). Otherwise, the curves are connected in a distorted figure eight (e.g., 2021 Dec 04). Bisecting the *eclipse begins/ends at sunrise and sunset* loops is the curve of maximum eclipse at sunrise (western loop) and sunset (eastern loop).

The eclipse magnitude is defined as the fraction of the Sun's diameter occulted by the Moon. A curve of constant eclipse magnitude delineates the points where the local magnitude at maximum eclipse is equal to a constant value. The maps include *curves of constant eclipse magnitude* for values of 0.2, 0.4, 0.6, and 0.8. These curves run exclusively between the curves of maximum eclipse at sunrise and sunset. They are approximately parallel to the northern/southern penumbral limits and the umbral/antumbral paths of central eclipses. The northern and southern limits of the penumbra may be thought of as curves of eclipse magnitude of 0.0. For total eclipses, the northern and southern limits of the umbra are curves of eclipse magnitude of 1.0.

Greatest eclipse is the instant when the axis of the Moon's shadow cone passes closest to Earth's center. The point on Earth's surface intersected by the axis of the Moon's shadow cone at greatest eclipse is marked by an asterisk. For partial eclipses, the shadow axis misses Earth entirely, so the point of greatest eclipse lies on the day/night terminator and the Sun appears on the horizon.

Parameters for the eclipse appears to the right of each map. They include the time of greatest eclipse (Universal Time[4] or UT1), the eclipse magnitude and obscuration, gamma[5], Saros series, and node (ascending or descending).

Photo 1–7 Time Sequence of the Total Solar Eclipse of 2017 Aug 21. ©2017 F. Espenak

---

[4] Universal Time (UT1) is the modern-day replacement for Greenwich Mean Time.
[5] Gamma is the distance of the Moon's shadow cone axis from Earth's center at greatest eclipse (in Earth equatorial radii).

Global Solar Eclipse Maps

Key to Global Solar Eclipse Maps

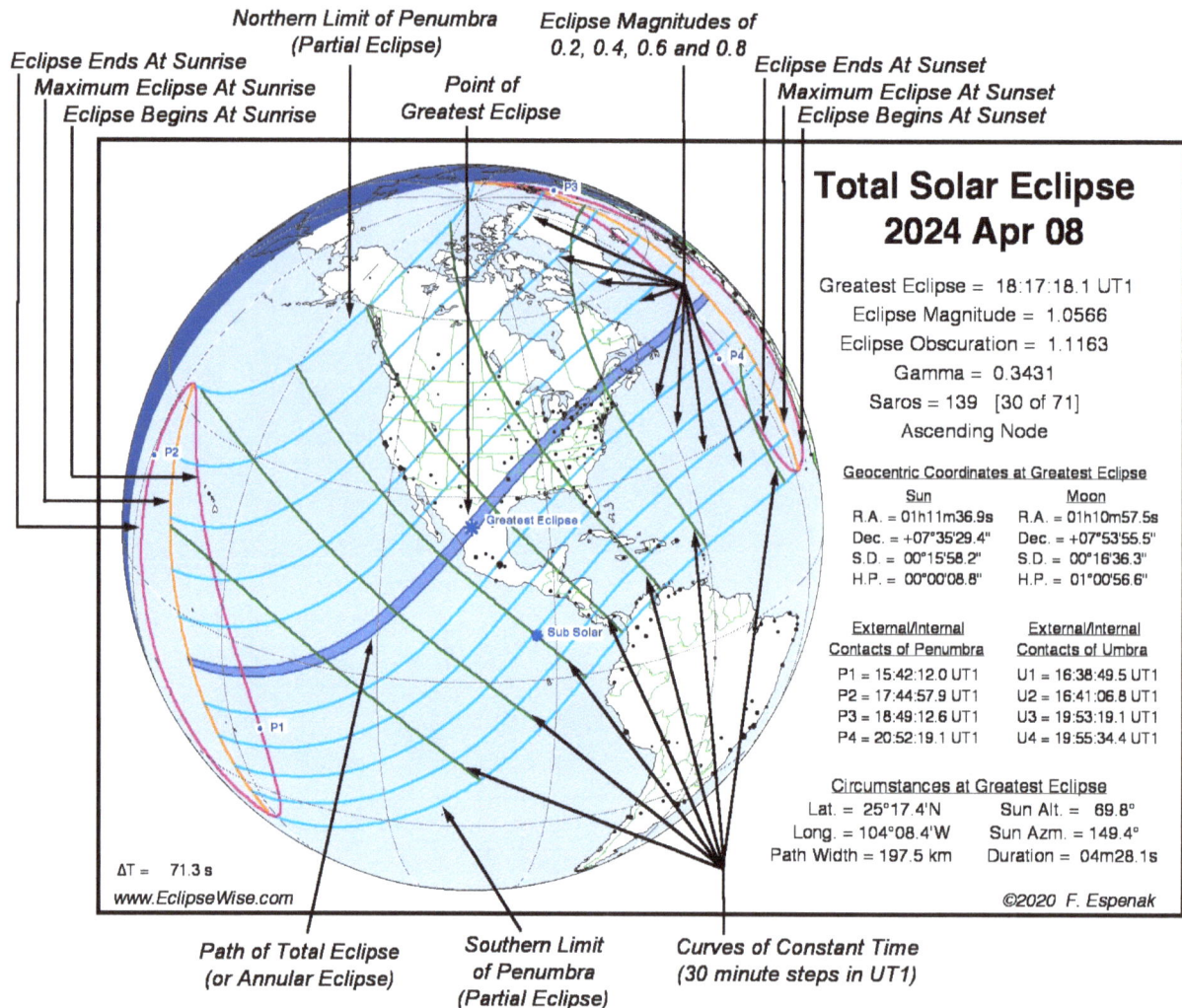

Total Solar Eclipse
2024 Apr 08

Greatest Eclipse = 18:17:18.1 UT1
Eclipse Magnitude = 1.0566
Eclipse Obscuration = 1.1163
Gamma = 0.3431
Saros = 139 [30 of 71]
Ascending Node

Geocentric Coordinates at Greatest Eclipse

Sun	Moon
R.A. = 01h11m36.9s	R.A. = 01h10m57.5s
Dec. = +07°35'29.4"	Dec. = +07°53'55.5"
S.D. = 00°15'58.2"	S.D. = 00°16'36.3"
H.P. = 00°00'08.8"	H.P. = 01°00'56.6"

External/Internal Contacts of Penumbra	External/Internal Contacts of Umbra
P1 = 15:42:12.0 UT1	U1 = 16:38:49.5 UT1
P2 = 17:44:57.9 UT1	U2 = 16:41:06.8 UT1
P3 = 18:49:12.6 UT1	U3 = 19:53:19.1 UT1
P4 = 20:52:19.1 UT1	U4 = 19:55:34.4 UT1

Circumstances at Greatest Eclipse

Lat. = 25°17.4'N	Sun Alt. = 69.8°
Long. = 104°08.4'W	Sun Azm. = 149.4°
Path Width = 197.5 km	Duration = 04m28.1s

ΔT = 71.3 s
www.EclipseWise.com
©2020 F. Espenak

Labels: Eclipse Ends At Sunrise / Maximum Eclipse At Sunrise / Eclipse Begins At Sunrise; Northern Limit of Penumbra (Partial Eclipse); Point of Greatest Eclipse; Eclipse Magnitudes of 0.2, 0.4, 0.6 and 0.8; Eclipse Ends At Sunset / Maximum Eclipse At Sunset / Eclipse Begins At Sunset; Path of Total Eclipse (or Annular Eclipse); Southern Limit of Penumbra (Partial Eclipse); Curves of Constant Time (30 minute steps in UT1)

Explanation of Terms Used in Global Solar Eclipse Maps

Greatest Eclipse – The instant when the distance between the axis of the Moon's shadow cone and the center of Earth reaches a minimum (in Universal Time[6] or UT1).

Eclipse Magnitude – The fraction of the Sun's diameter occulted by the Moon at the instant of greatest eclipse (for total and annular eclipses this value is the ratio of diameters of the Moon and the Sun).

Eclipse Obscuration – The fraction of the Sun's area occulted by the Moon at the instant of greatest eclipse.

Gamma – The minimum distance from the lunar shadow axis to the center of Earth (units of Earth equatorial radii).

Saros Series – The Saros series that the eclipse belongs to. The numbers in "[ ]" are the eclipse's sequential position and the number of eclipses in the Saros series.

Node – The orbital node near which the eclipse takes place (Ascending Node or Descending Node).

Geocentric Coordinates of the Sun and the Moon at Greatest Eclipse

        R.A. – Right Ascension   **S.D.** – Semi-Diameter (i.e. - radius)
        Dec. – Declination   **H.P.** – Horizontal Parallax

External/Internal Contacts of Penumbra and Umbra – Instants when each shadow enters or exits the surface of Earth (Penumbral contacts shown on map as: P1, P2, P3, and P4; umbral contacts are located at the ends of the central path). All times are in Universal Time (UT1)

Circumstances of Greatest Eclipse – Geographic location (Lat., Long.) of the shadow axis, the altitude and azimuth of the Sun, width of the central path, and the central duration of totality or annularity.

---

[6] Universal Time or UT1 is the modern replacement for Greenwich Mean Time

Total Solar Eclipse
2061 Apr 20

Greatest Eclipse = 02:55:16.2 UT1

Eclipse Magnitude = 1.0476

Eclipse Obscuration = 1.0974

Gamma = 0.9578

Saros = 149 [23 of 71]

Ascending Node

Geocentric Coordinates at Greatest Eclipse

	Sun	Moon
R.A. =	01h53m47.8s	R.A. = 01h52m03.2s
Dec. =	+11°39'59.8"	Dec. = +12°32'19.1"
S.D. =	00°15'55.3"	S.D. = 00°16'36.4"
H.P. =	00°00'08.8"	H.P. = 01°00'56.9"

External/Internal Contacts of Penumbra

P1 = 00:51:00.0 UT1
P4 = 04:59:10.3 UT1

External/Internal Contacts of Umbra

U1 = 02:22:14.3 UT1
U2 = 02:30:33.3 UT1
U3 = 03:19:27.3 UT1
U4 = 03:27:49.4 UT1

Circumstances at Greatest Eclipse

Lat. = 64°32.2'N	Sun Alt. = 16.2°
Long. = 059°03.7'E	Sun Azm. = 96.8°
Path Width = 558.8 km	Duration = 02m37.4s

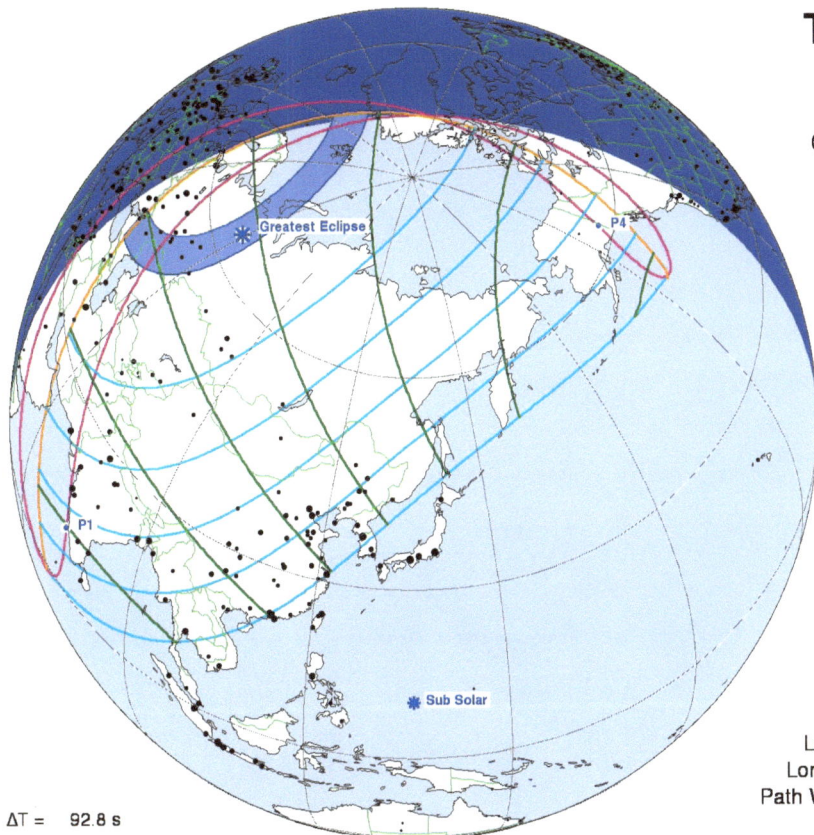

ΔT = 92.8 s

www.EclipseWise.com

©2020 F. Espenak

Annular Solar Eclipse
2061 Oct 13

Greatest Eclipse = 10:30:36.5 UT1

Eclipse Magnitude = 0.9469

Eclipse Obscuration = 0.8966

Gamma = -0.9639

Saros = 154 [9 of 71]

Decending Node

Geocentric Coordinates at Greatest Eclipse

	Sun	Moon
R.A. =	13h16m11.1s	R.A. = 13h14m30.5s
Dec. =	-08°03'03.6"	Dec. = -08°50'16.1"
S.D. =	00°16'01.7"	S.D. = 00°15'07.5"
H.P. =	00°00'08.8"	H.P. = 00°55'30.4"

External/Internal Contacts of Penumbra

P1 = 08:09:03.8 UT1
P4 = 12:51:51.5 UT1

External/Internal Contacts of Umbra

U1 = 09:54:03.9 UT1
U2 = 10:09:18.6 UT1
U3 = 10:51:14.8 UT1
U4 = 11:06:36.3 UT1

Circumstances at Greatest Eclipse

Lat. = 62°08.2'S	Sun Alt. = 14.9°
Long. = 054°28.8'W	Sun Azm. = 79.0°
Path Width = 742.6 km	Duration = 03m41.0s

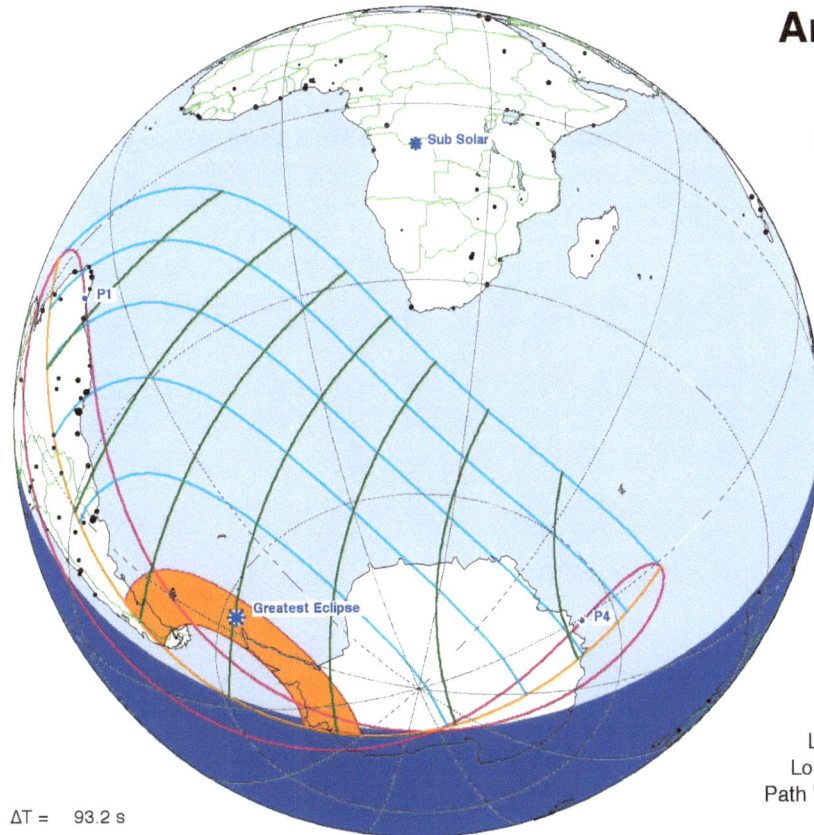

ΔT = 93.2 s

www.EclipseWise.com

©2020 F. Espenak

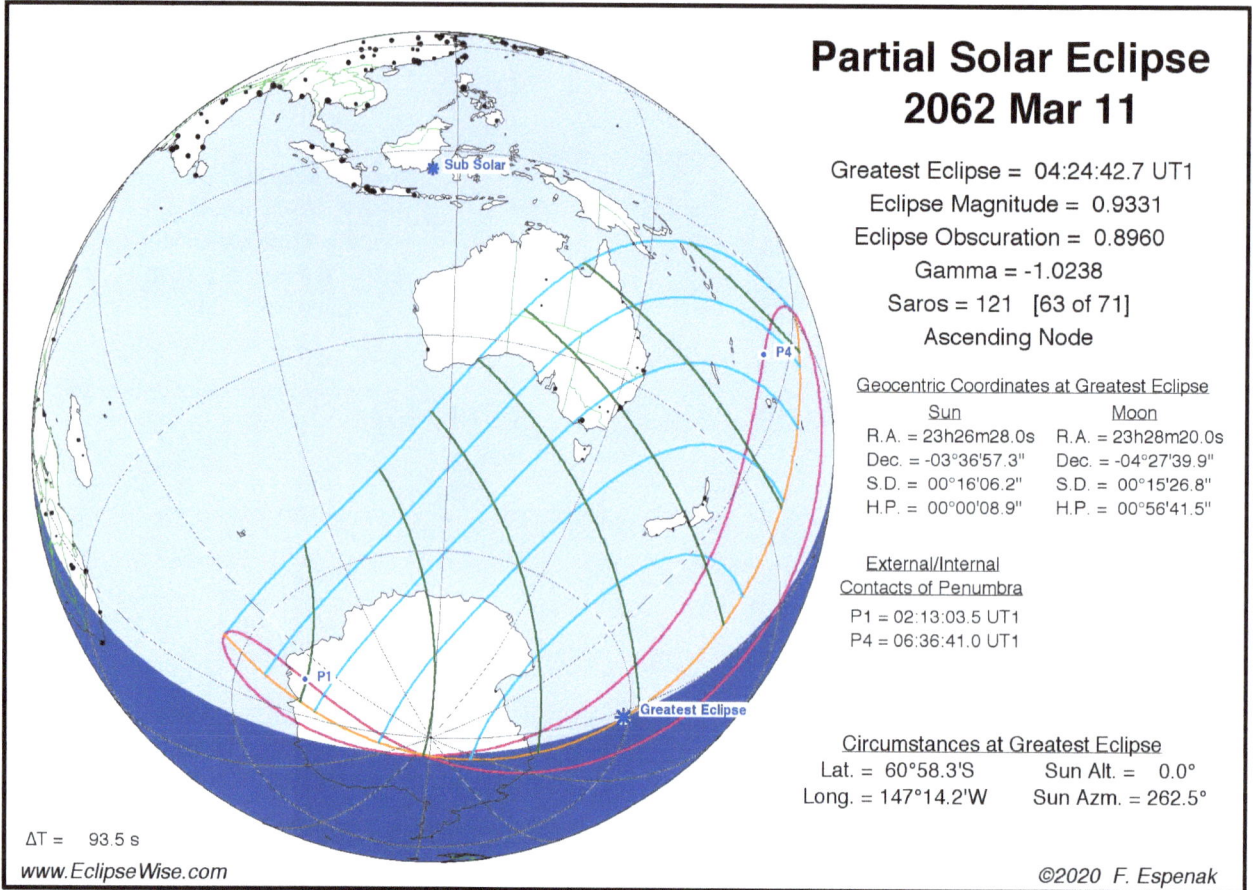

Partial Solar Eclipse
2062 Mar 11

Greatest Eclipse = 04:24:42.7 UT1
Eclipse Magnitude = 0.9331
Eclipse Obscuration = 0.8960
Gamma = -1.0238
Saros = 121   [63 of 71]
Ascending Node

Geocentric Coordinates at Greatest Eclipse

	Sun	Moon
R.A. =	23h26m28.0s	R.A. = 23h28m20.0s
Dec. =	-03°36'57.3"	Dec. = -04°27'39.9"
S.D. =	00°16'06.2"	S.D. = 00°15'26.8"
H.P. =	00°00'08.9"	H.P. = 00°56'41.5"

External/Internal Contacts of Penumbra

P1 = 02:13:03.5 UT1
P4 = 06:36:41.0 UT1

Circumstances at Greatest Eclipse

Lat. = 60°58.3'S	Sun Alt. = 0.0°
Long. = 147°14.2'W	Sun Azm. = 262.5°

ΔT = 93.5 s

www.EclipseWise.com

©2020 F. Espenak

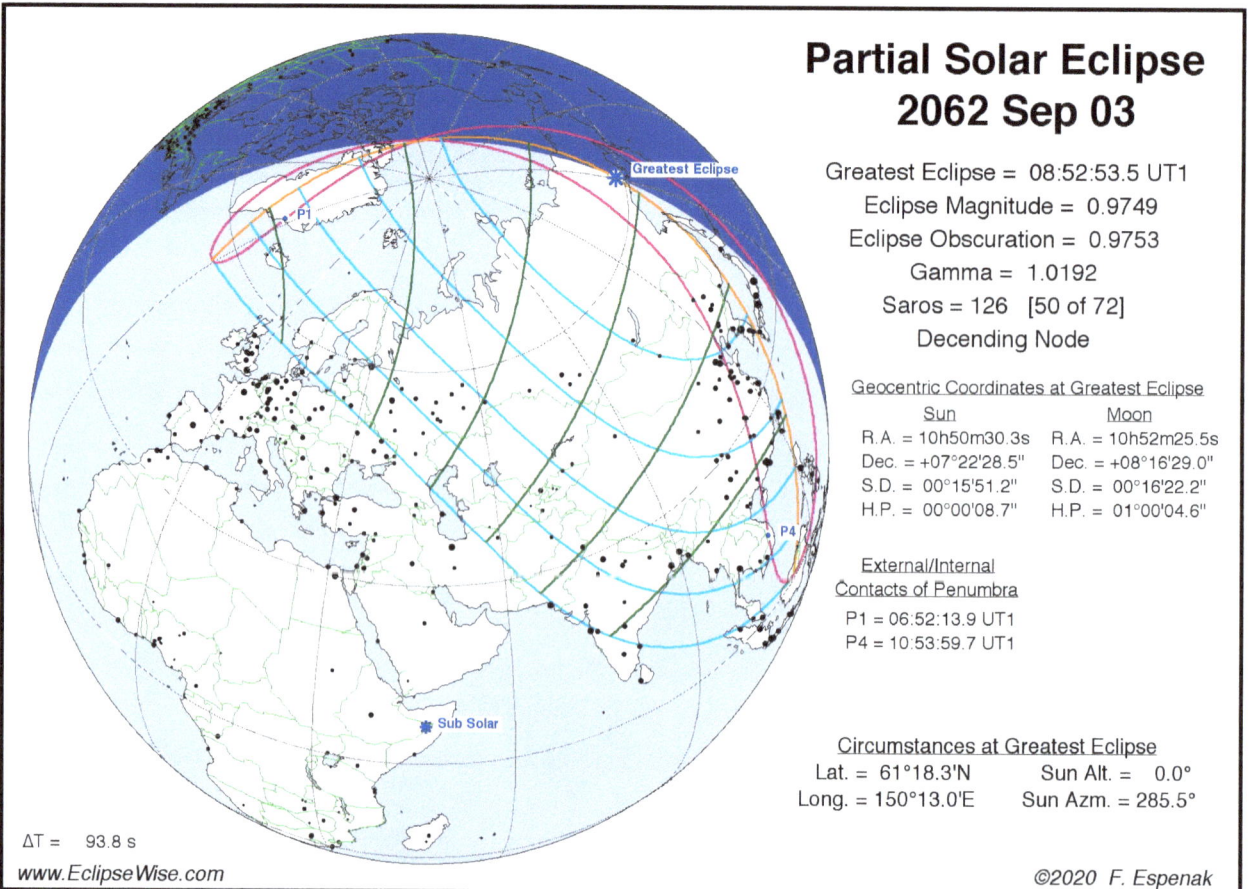

Partial Solar Eclipse
2062 Sep 03

Greatest Eclipse = 08:52:53.5 UT1
Eclipse Magnitude = 0.9749
Eclipse Obscuration = 0.9753
Gamma = 1.0192
Saros = 126   [50 of 72]
Decending Node

Geocentric Coordinates at Greatest Eclipse

	Sun	Moon
R.A. =	10h50m30.3s	R.A. = 10h52m25.5s
Dec. =	+07°22'28.5"	Dec. = +08°16'29.0"
S.D. =	00°15'51.2"	S.D. = 00°16'22.2"
H.P. =	00°00'08.7"	H.P. = 01°00'04.6"

External/Internal Contacts of Penumbra

P1 = 06:52:13.9 UT1
P4 = 10:53:59.7 UT1

Circumstances at Greatest Eclipse

Lat. = 61°18.3'N	Sun Alt. = 0.0°
Long. = 150°13.0'E	Sun Azm. = 285.5°

ΔT = 93.8 s

www.EclipseWise.com

©2020 F. Espenak

Annular Solar Eclipse
2063 Feb 28

Greatest Eclipse = 07:41:55.8 UT1

Eclipse Magnitude = 0.9293

Eclipse Obscuration = 0.8635

Gamma = -0.3360

Saros = 131 [53 of 70]

Ascending Node

Geocentric Coordinates at Greatest Eclipse

	Sun	Moon
R.A. =	22h45m11.8s	22h45m46.2s
Dec. =	-07°54'42.4"	-08°10'47.1"
S.D. =	00°16'08.9"	00°14'47.6"
H.P. =	00°00'08.9"	00°54'17.7"

External/Internal Contacts of Penumbra	External/Internal Contacts of Umbra
P1 = 04:40:31.4 UT1	U1 = 05:47:35.8 UT1
P2 = 07:11:06.3 UT1	U2 = 05:53:57.4 UT1
P3 = 08:13:16.4 UT1	U3 = 09:30:08.7 UT1
P4 = 10:43:24.8 UT1	U4 = 09:36:26.4 UT1

Circumstances at Greatest Eclipse

Lat. = 25°13.9'S	Sun Alt. = 70.2°
Long. = 077°38.2'E	Sun Azm. = 329.4°
Path Width = 279.6 km	Duration = 07m41.2s

ΔT = 94.2 s

www.EclipseWise.com

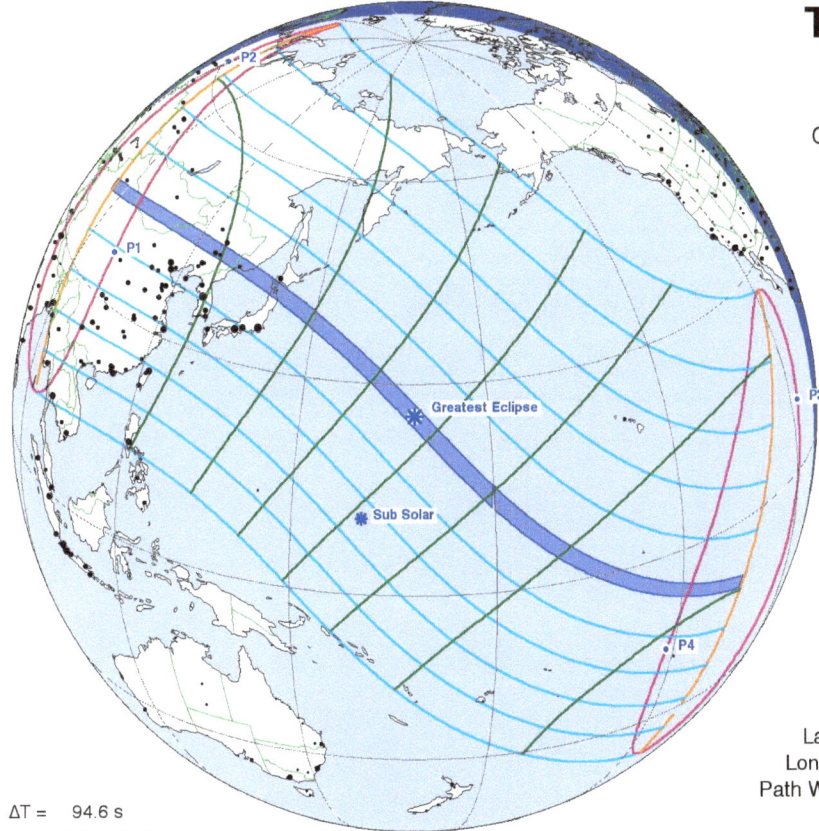

©2020 F. Espenak

Total Solar Eclipse
2063 Aug 24

Greatest Eclipse = 01:20:36.0 UT1

Eclipse Magnitude = 1.0750

Eclipse Obscuration = 1.1556

Gamma = 0.2771

Saros = 136 [40 of 71]

Decending Node

Geocentric Coordinates at Greatest Eclipse

	Sun	Moon
R.A. =	10h12m03.7s	10h12m34.5s
Dec. =	+11°07'34.9"	+11°22'46.8"
S.D. =	00°15'48.9"	00°16'43.4"
H.P. =	00°00'08.7"	01°01'22.6"

External/Internal Contacts of Penumbra	External/Internal Contacts of Umbra
P1 = 22:46:00.1 UT1	U1 = 23:40:29.5 UT1
P2 = 00:42:07.2 UT1	U2 = 23:43:37.7 UT1
P3 = 01:59:23.6 UT1	U3 = 02:57:43.1 UT1
P4 = 03:55:17.5 UT1	U4 = 03:00:51.2 UT1

Circumstances at Greatest Eclipse

Lat. = 25°33.7'N	Sun Alt. = 73.8°
Long. = 168°19.6'E	Sun Azm. = 208.6°
Path Width = 252.2 km	Duration = 05m49.1s

ΔT = 94.6 s

www.EclipseWise.com

©2020 F. Espenak

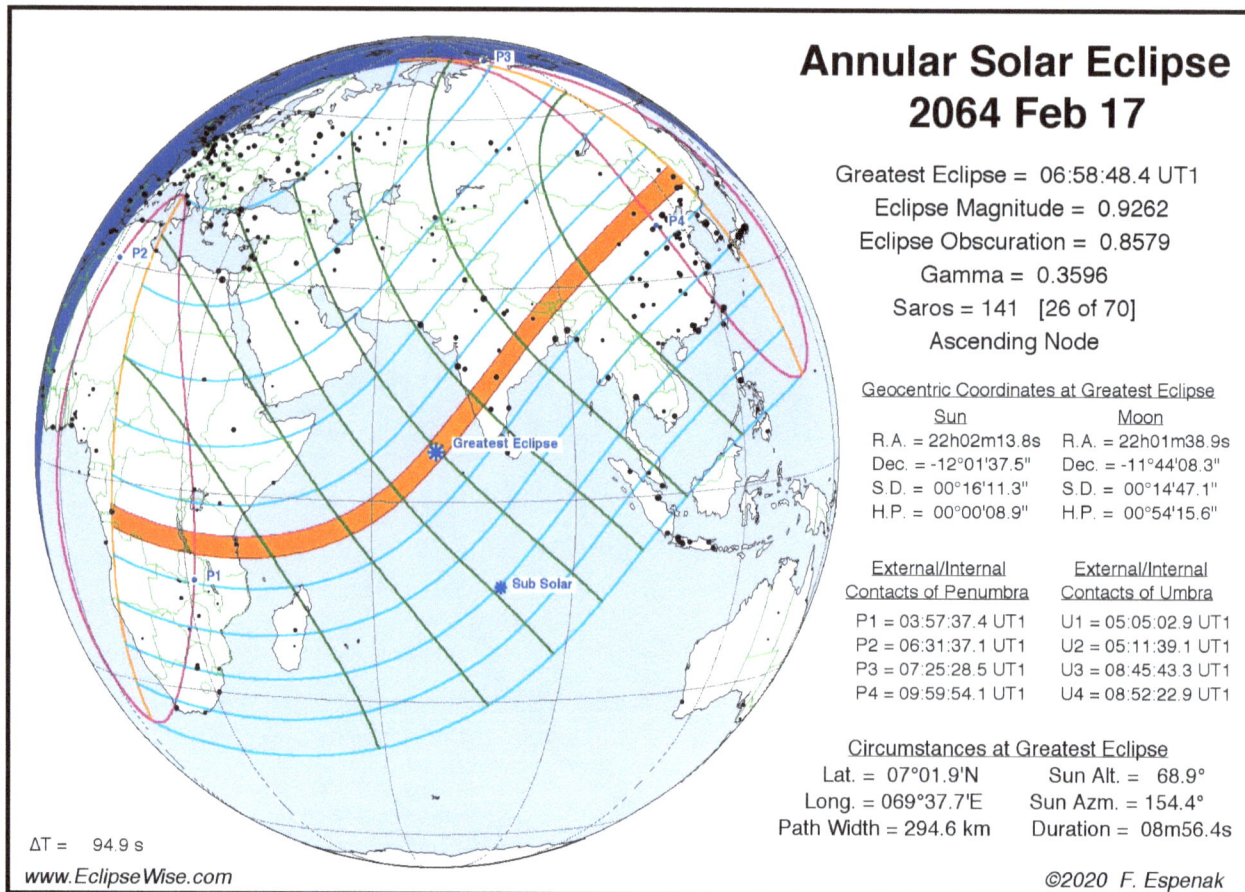

Annular Solar Eclipse
2064 Feb 17

Greatest Eclipse = 06:58:48.4 UT1

Eclipse Magnitude = 0.9262

Eclipse Obscuration = 0.8579

Gamma = 0.3596

Saros = 141 [26 of 70]

Ascending Node

Geocentric Coordinates at Greatest Eclipse

Sun	Moon
R.A. = 22h02m13.8s	R.A. = 22h01m38.9s
Dec. = -12°01'37.5"	Dec. = -11°44'08.3"
S.D. = 00°16'11.3"	S.D. = 00°14'47.1"
H.P. = 00°00'08.9"	H.P. = 00°54'15.6"

External/Internal Contacts of Penumbra	External/Internal Contacts of Umbra
P1 = 03:57:37.4 UT1	U1 = 05:05:02.9 UT1
P2 = 06:31:37.1 UT1	U2 = 05:11:39.1 UT1
P3 = 07:25:28.5 UT1	U3 = 08:45:43.3 UT1
P4 = 09:59:54.1 UT1	U4 = 08:52:22.9 UT1

Circumstances at Greatest Eclipse

Lat. = 07°01.9'N	Sun Alt. = 68.9°
Long. = 069°37.7'E	Sun Azm. = 154.4°
Path Width = 294.6 km	Duration = 08m56.4s

ΔT = 94.9 s

www.EclipseWise.com

©2020 F. Espenak

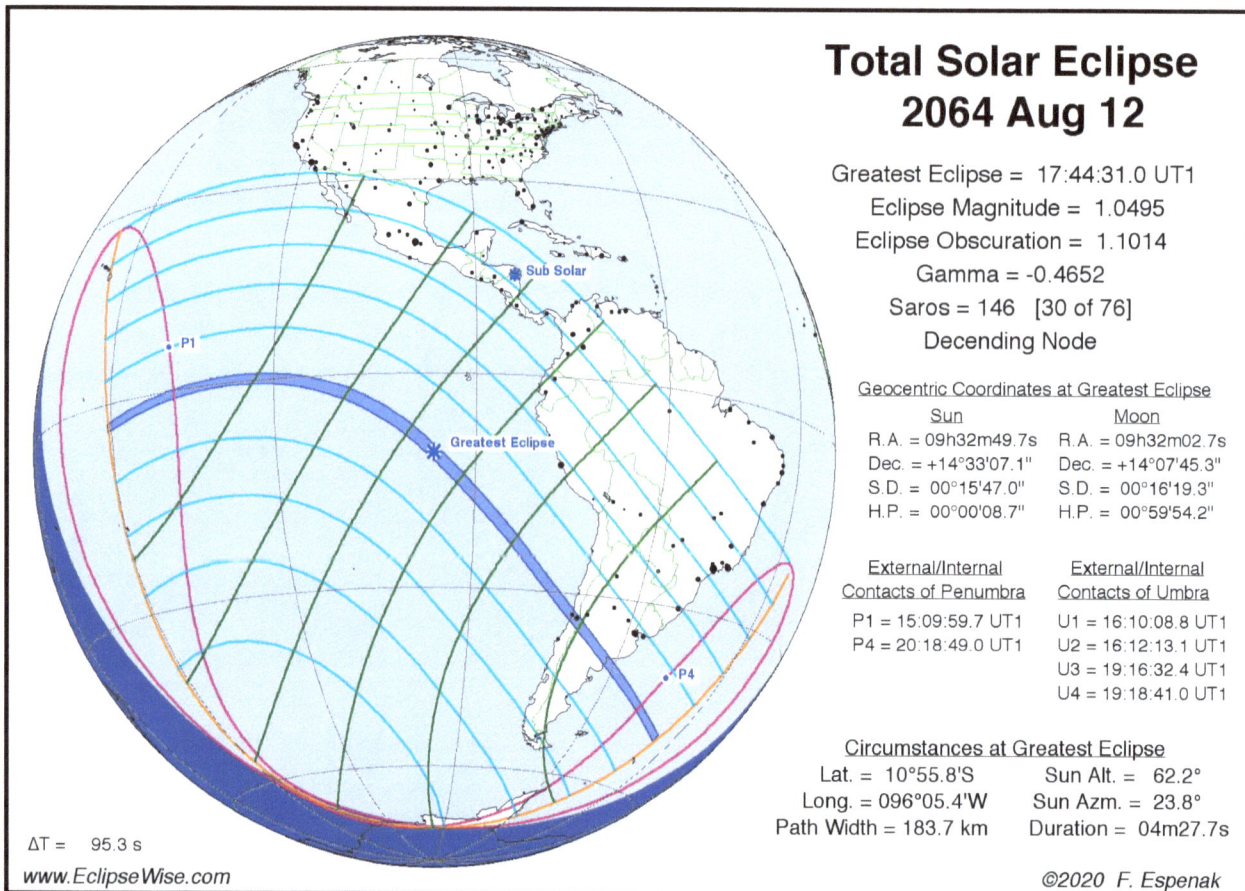

Total Solar Eclipse
2064 Aug 12

Greatest Eclipse = 17:44:31.0 UT1

Eclipse Magnitude = 1.0495

Eclipse Obscuration = 1.1014

Gamma = -0.4652

Saros = 146 [30 of 76]

Decending Node

Geocentric Coordinates at Greatest Eclipse

Sun	Moon
R.A. = 09h32m49.7s	R.A. = 09h32m02.7s
Dec. = +14°33'07.1"	Dec. = +14°07'45.3"
S.D. = 00°15'47.0"	S.D. = 00°16'19.3"
H.P. = 00°00'08.7"	H.P. = 00°59'54.2"

External/Internal Contacts of Penumbra	External/Internal Contacts of Umbra
P1 = 15:09:59.7 UT1	U1 = 16:10:08.8 UT1
P4 = 20:18:49.0 UT1	U2 = 16:12:13.1 UT1
	U3 = 19:16:32.4 UT1
	U4 = 19:18:41.0 UT1

Circumstances at Greatest Eclipse

Lat. = 10°55.8'S	Sun Alt. = 62.2°
Long. = 096°05.4'W	Sun Azm. = 23.8°
Path Width = 183.7 km	Duration = 04m27.7s

ΔT = 95.3 s

www.EclipseWise.com

©2020 F. Espenak

Partial Solar Eclipse
2065 Feb 05

Greatest Eclipse = 09:50:49.9 UT1

Eclipse Magnitude = 0.9123

Eclipse Obscuration = 0.8676

Gamma = 1.0336

Saros = 151 [17 of 72]

Ascending Node

Geocentric Coordinates at Greatest Eclipse

	Sun	Moon
R.A. =	21h18m22.7s	21h16m47.2s
Dec. =	-15°41'30.6"	-14°47'52.5"
S.D. =	00°16'13.3"	00°15'25.7"
H.P. =	00°00'08.9"	00°56'37.5"

External/Internal
Contacts of Penumbra

P1 = 07:39:09.6 UT1

P4 = 12:02:15.6 UT1

Circumstances at Greatest Eclipse

Lat. = 62°09.7'N	Sun Alt. = 0.0°
Long. = 022°01.0'W	Sun Azm. = 125.4°

ΔT =   95.7 s

www.EclipseWise.com

©2020 F. Espenak

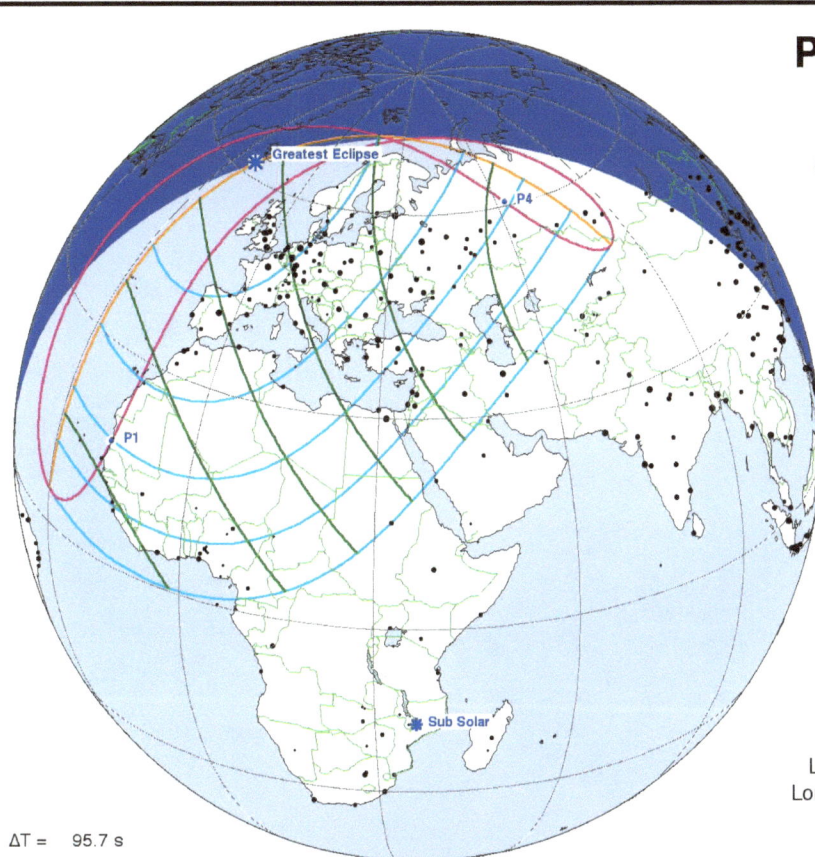

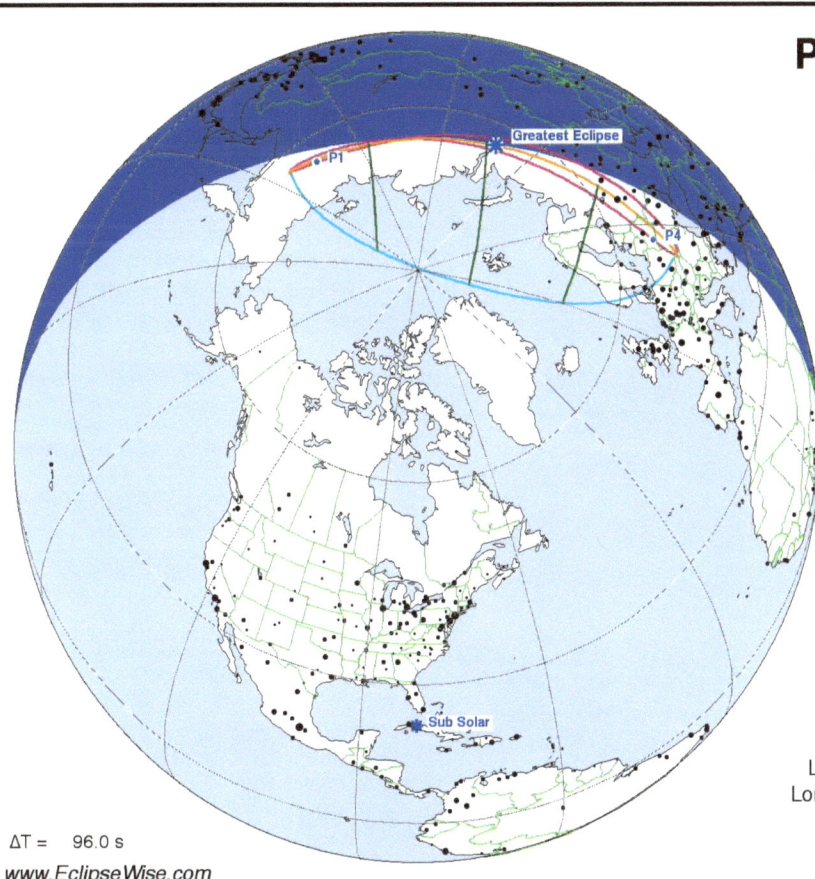

Partial Solar Eclipse
2065 Jul 03

Greatest Eclipse = 17:32:16.5 UT1

Eclipse Magnitude = 0.1639

Eclipse Obscuration = 0.0768

Gamma = 1.4619

Saros = 118 [71 of 72]

Decending Node

Geocentric Coordinates at Greatest Eclipse

	Sun	Moon
R.A. =	06h53m43.9s	06h54m50.6s
Dec. =	+22°51'26.7"	+24°10'43.8"
S.D. =	00°15'43.9"	00°15'05.3"
H.P. =	00°00'08.6"	00°55'22.6"

External/Internal
Contacts of Penumbra

P1 = 16:31:08.5 UT1

P4 = 18:33:34.0 UT1

Circumstances at Greatest Eclipse

Lat. = 64°49.3'N	Sun Alt. = 0.0°
Long. = 071°44.3'E	Sun Azm. = 335.9°

ΔT =   96.0 s

www.EclipseWise.com

©2020 F. Espenak

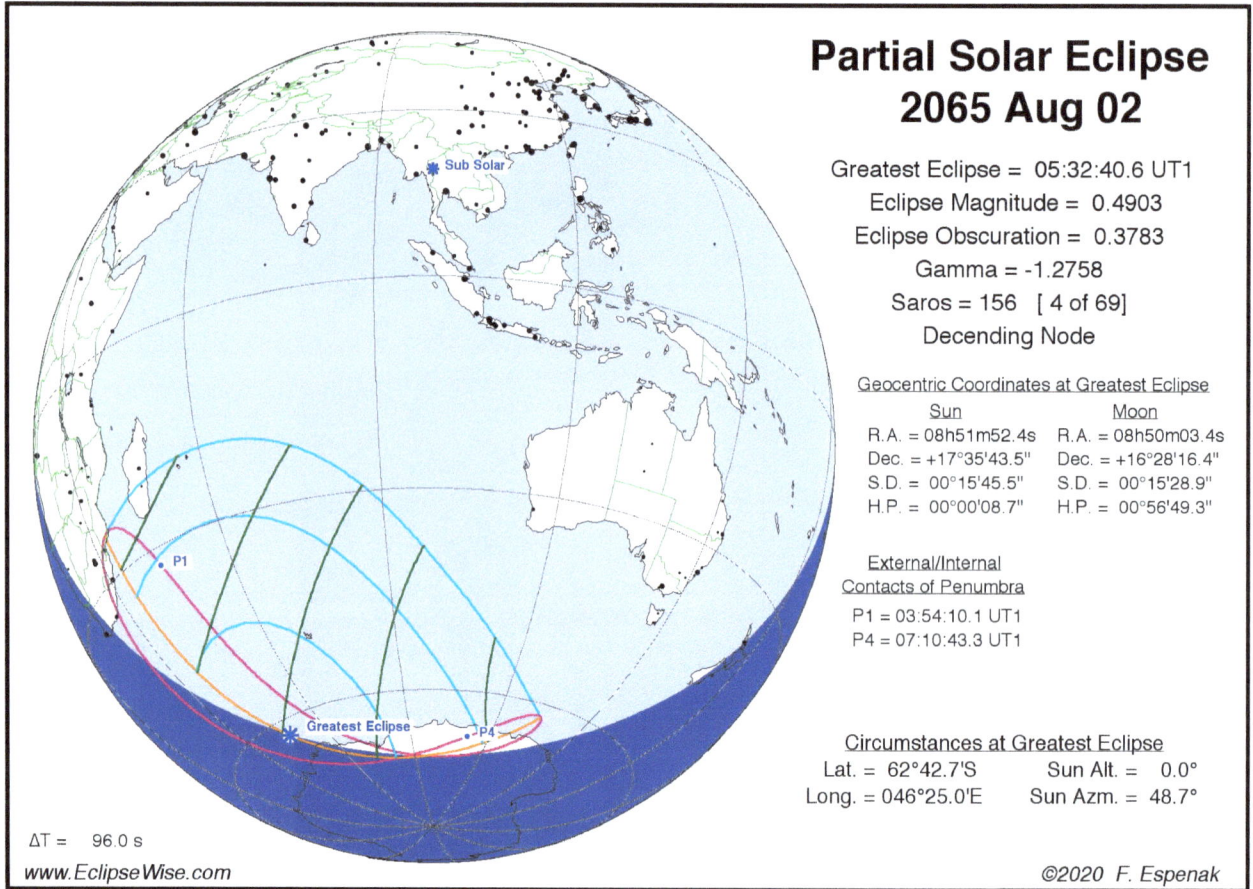

Partial Solar Eclipse
2065 Aug 02

Greatest Eclipse = 05:32:40.6 UT1

Eclipse Magnitude = 0.4903

Eclipse Obscuration = 0.3783

Gamma = -1.2758

Saros = 156 [4 of 69]

Decending Node

Geocentric Coordinates at Greatest Eclipse

	Sun	Moon
R.A. =	08h51m52.4s	08h50m03.4s
Dec. =	+17°35'43.5"	+16°28'16.4"
S.D. =	00°15'45.5"	00°15'28.9"
H.P. =	00°00'08.7"	00°56'49.3"

External/Internal
Contacts of Penumbra

P1 = 03:54:10.1 UT1

P4 = 07:10:43.3 UT1

Circumstances at Greatest Eclipse

Lat. = 62°42.7'S	Sun Alt. = 0.0°
Long. = 046°25.0'E	Sun Azm. = 48.7°

ΔT = 96.0 s

www.EclipseWise.com

©2020 F. Espenak

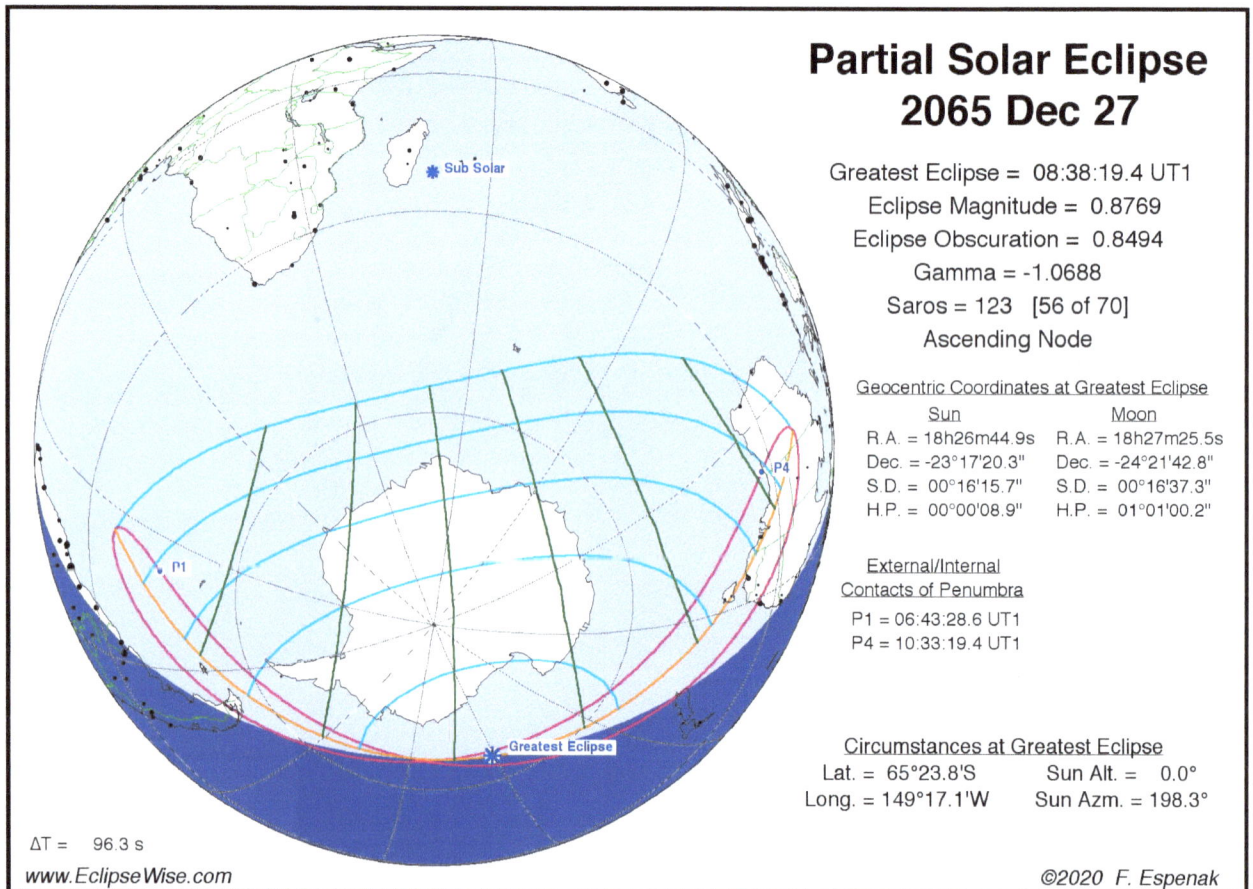

Partial Solar Eclipse
2065 Dec 27

Greatest Eclipse = 08:38:19.4 UT1

Eclipse Magnitude = 0.8769

Eclipse Obscuration = 0.8494

Gamma = -1.0688

Saros = 123 [56 of 70]

Ascending Node

Geocentric Coordinates at Greatest Eclipse

	Sun	Moon
R.A. =	18h26m44.9s	18h27m25.5s
Dec. =	-23°17'20.3"	-24°21'42.8"
S.D. =	00°16'15.7"	00°16'37.3"
H.P. =	00°00'08.9"	01°01'00.2"

External/Internal
Contacts of Penumbra

P1 = 06:43:28.6 UT1

P4 = 10:33:19.4 UT1

Circumstances at Greatest Eclipse

Lat. = 65°23.8'S	Sun Alt. = 0.0°
Long. = 149°17.1'W	Sun Azm. = 198.3°

ΔT = 96.3 s

www.EclipseWise.com

©2020 F. Espenak

Annular Solar Eclipse
2066 Jun 22

Greatest Eclipse = 19:24:11.0 UT1

Eclipse Magnitude = 0.9435

Eclipse Obscuration = 0.8901

Gamma = 0.7330

Saros = 128 [61 of 73]

Decending Node

Geocentric Coordinates at Greatest Eclipse

Sun	Moon
R.A. = 06h07m28.7s	R.A. = 06h07m48.1s
Dec. = +23°25'11.2"	Dec. = +24°04'22.4"
S.D. = 00°15'44.2"	S.D. = 00°14'42.0"
H.P. = 00°00'08.7"	H.P. = 00°53'57.0"

External/Internal Contacts of Penumbra	External/Internal Contacts of Umbra
P1 = 16:40:06.4 UT1	U1 = 18:00:24.0 UT1
P4 = 22:08:19.3 UT1	U2 = 18:07:13.5 UT1
	U3 = 20:41:15.3 UT1
	U4 = 20:48:04.1 UT1

Circumstances at Greatest Eclipse

Lat. = 70°07.9'N	Sun Alt. = 42.6°
Long. = 096°29.4'W	Sun Azm. = 197.5°
Path Width = 308.5 km	Duration = 04m39.8s

ΔT = 96.7 s

www.EclipseWise.com

©2020 F. Espenak

Total Solar Eclipse
2066 Dec 17

Greatest Eclipse = 00:22:02.8 UT1

Eclipse Magnitude = 1.0416

Eclipse Obscuration = 1.0848

Gamma = -0.4043

Saros = 133 [48 of 72]

Ascending Node

Geocentric Coordinates at Greatest Eclipse

Sun	Moon
R.A. = 17h39m46.4s	R.A. = 17h39m53.3s
Dec. = -23°20'56.0"	Dec. = -23°45'32.9"
S.D. = 00°16'15.1"	S.D. = 00°16'39.9"
H.P. = 00°00'08.9"	H.P. = 01°01'09.6"

External/Internal Contacts of Penumbra	External/Internal Contacts of Umbra
P1 = 21:48:21.2 UT1	U1 = 22:46:44.2 UT1
P2 = 23:59:45.5 UT1	U2 = 22:48:13.1 UT1
P3 = 00:44:24.6 UT1	U3 = 01:55:53.3 UT1
P4 = 02:55:43.5 UT1	U4 = 01:57:24.1 UT1

Circumstances at Greatest Eclipse

Lat. = 47°22.1'S	Sun Alt. = 65.9°
Long. = 175°38.8'E	Sun Azm. = 355.1°
Path Width = 152.2 km	Duration = 03m14.5s

ΔT = 97.1 s

www.EclipseWise.com

©2020 F. Espenak

Annular Solar Eclipse
2067 Jun 11

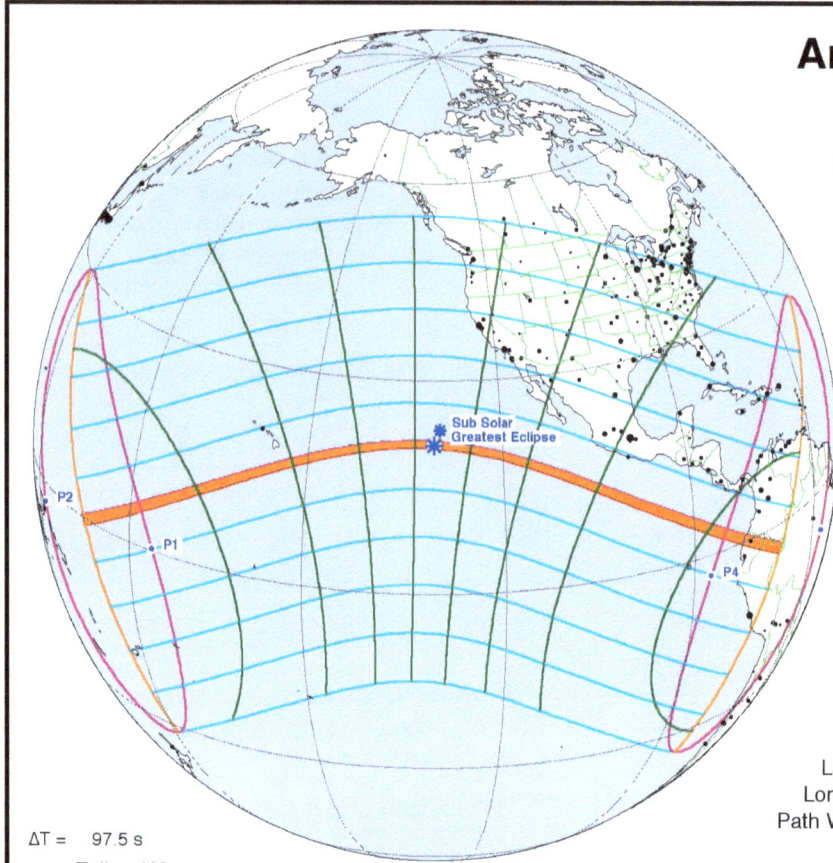

Greatest Eclipse = 20:40:49.0 UT1

Eclipse Magnitude = 0.9670

Eclipse Obscuration = 0.9351

Gamma = -0.0387

Saros = 138   [34 of 70]

Decending Node

Geocentric Coordinates at Greatest Eclipse

	Sun	Moon
R.A. =	05h20m58.3s	R.A. = 05h20m58.0s
Dec. =	+23°07'36.6"	Dec. = +23°05'29.3"
S.D. =	00°15'45.1"	S.D. = 00°15'00.0"
H.P. =	00°00'08.7"	H.P. = 00°55'03.2"

External/Internal Contacts of Penumbra	External/Internal Contacts of Umbra
P1 = 17:40:04.6 UT1	U1 = 18:43:24.5 UT1
P2 = 19:50:01.2 UT1	U2 = 18:46:35.1 UT1
P3 = 21:31:35.6 UT1	U3 = 22:35:01.1 UT1
P4 = 23:41:38.4 UT1	U4 = 22:38:16.3 UT1

Circumstances at Greatest Eclipse

Lat. = 21°02.4'N	Sun Alt. = 87.9°
Long. = 130°18.7'W	Sun Azm. = 2.0°
Path Width = 118.9 km	Duration = 04m05.2s

ΔT =   97.5 s

www.EclipseWise.com

©2020 F. Espenak

Hybrid Solar Eclipse
2067 Dec 06

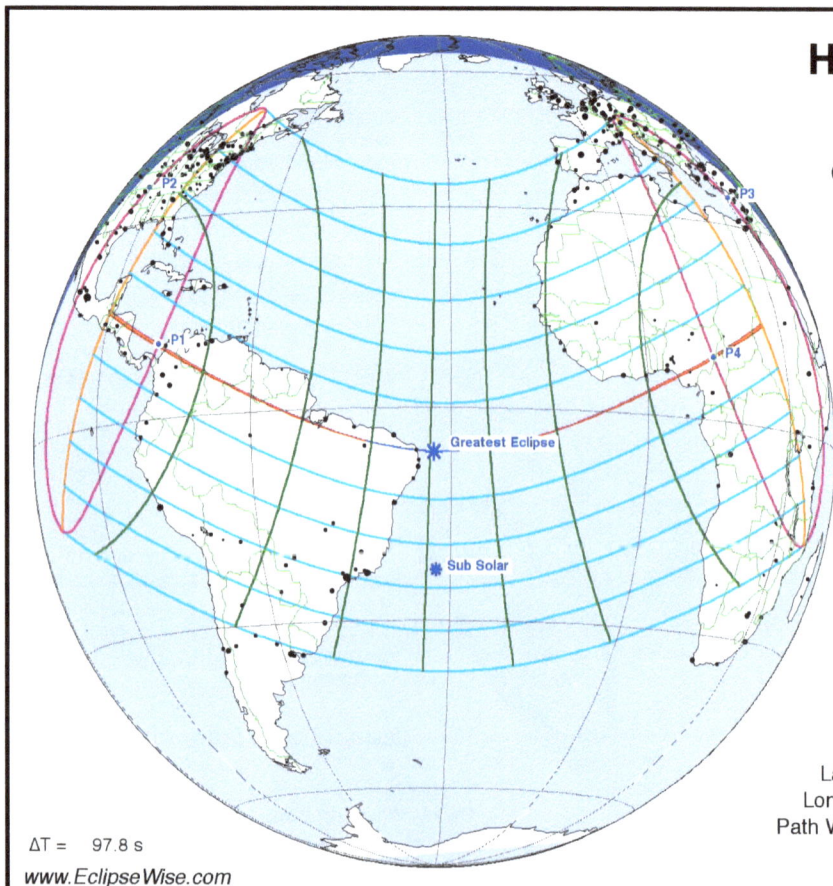

Greatest Eclipse = 14:02:05.4 UT1

Eclipse Magnitude = 1.0011

Eclipse Obscuration = 1.0023

Gamma = 0.2845

Saros = 143   [26 of 72]

Ascending Node

Geocentric Coordinates at Greatest Eclipse

	Sun	Moon
R.A. =	16h52m45.8s	R.A. = 16h52m46.9s
Dec. =	-22°31'49.1"	Dec. = -22°15'09.9"
S.D. =	00°16'13.8"	S.D. = 00°15'59.7"
H.P. =	00°00'08.9"	H.P. = 00°58'42.2"

External/Internal Contacts of Penumbra	External/Internal Contacts of Umbra
P1 = 11:17:07.7 UT1	U1 = 12:17:52.6 UT1
P2 = 13:24:35.5 UT1	U2 = 12:18:52.0 UT1
P3 = 14:39:37.2 UT1	U3 = 15:45:21.2 UT1
P4 = 16:46:57.5 UT1	U4 = 15:46:15.2 UT1

Circumstances at Greatest Eclipse

Lat. = 06°02.8'S	Sun Alt. = 73.5°
Long. = 032°30.7'W	Sun Azm. = 180.9°
Path Width = 4.1 km	Duration = 00m07.5s

ΔT =   97.8 s

www.EclipseWise.com

©2020 F. Espenak

Total Solar Eclipse
2068 May 31

Greatest Eclipse = 03:55:00.8 UT1

Eclipse Magnitude = 1.0110

Eclipse Obscuration = 1.0221

Gamma = -0.7970

Saros = 148 [24 of 75]

Decending Node

Geocentric Coordinates at Greatest Eclipse

Sun	Moon
R.A. = 04h35m49.8s	R.A. = 04h35m58.7s
Dec. = +22°01'13.9"	Dec. = +21°15'11.0"
S.D. = 00°15'46.5"	S.D. = 00°15'47.8"
H.P. = 00°00'08.7"	H.P. = 00°57'58.6"

External/Internal Contacts of Penumbra	External/Internal Contacts of Umbra
P1 = 01:30:22.5 UT1	U1 = 02:49:17.6 UT1
P4 = 06:19:46.6 UT1	U2 = 02:49:25.1 UT1
	U3 = 05:00:42.3 UT1
	U4 = 05:00:44.3 UT1

Circumstances at Greatest Eclipse

Lat. = 31°00.8'S	Sun Alt. = 36.9°
Long. = 123°05.5'E	Sun Azm. = 357.2°
Path Width = 62.8 km	Duration = 01m05.8s

ΔT = 98.2 s

www.EclipseWise.com

©2020 F. Espenak

Partial Solar Eclipse
2068 Nov 24

Greatest Eclipse = 21:30:51.0 UT1

Eclipse Magnitude = 0.9109

Eclipse Obscuration = 0.8547

Gamma = 1.0299

Saros = 153 [12 of 70]

Ascending Node

Geocentric Coordinates at Greatest Eclipse

Sun	Moon
R.A. = 16h05m39.1s	R.A. = 16h06m01.8s
Dec. = -20°49'55.6"	Dec. = -19°53'06.5"
S.D. = 00°16'12.0"	S.D. = 00°15'08.3"
H.P. = 00°00'08.9"	H.P. = 00°55'33.5"

External/Internal Contacts of Penumbra

P1 = 19:15:13.5 UT1

P4 = 23:46:28.4 UT1

Circumstances at Greatest Eclipse

Lat. = 68°31.0'N	Sun Alt. = 0.0°
Long. = 131°13.0'W	Sun Azm. = 193.8°

ΔT = 98.6 s

www.EclipseWise.com

©2020 F. Espenak

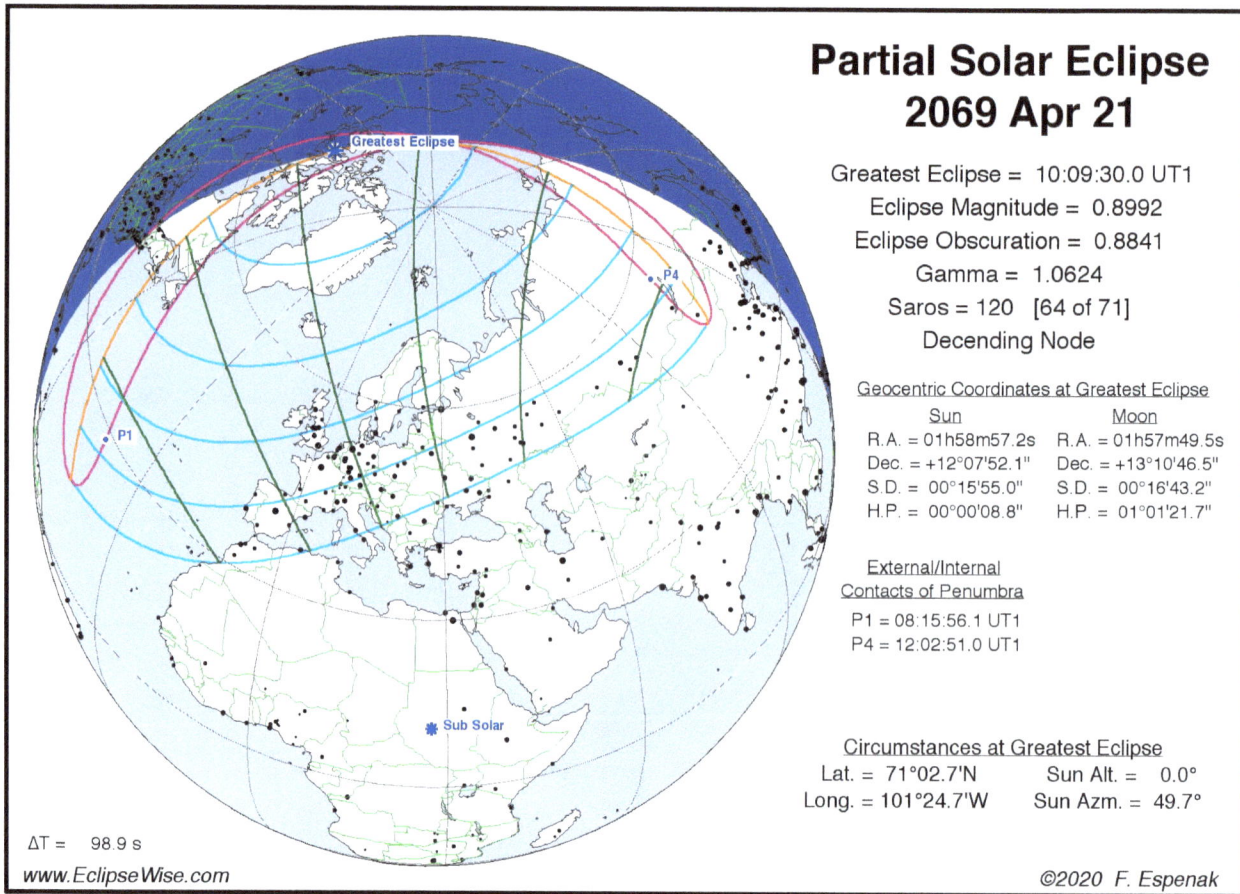

Partial Solar Eclipse
2069 Apr 21

Greatest Eclipse = 10:09:30.0 UT1
Eclipse Magnitude = 0.8992
Eclipse Obscuration = 0.8841
Gamma = 1.0624
Saros = 120 [64 of 71]
Decending Node

Geocentric Coordinates at Greatest Eclipse
Sun	Moon
R.A. = 01h58m57.2s	R.A. = 01h57m49.5s
Dec. = +12°07'52.1"	Dec. = +13°10'46.5"
S.D. = 00°15'55.0"	S.D. = 00°16'43.2"
H.P. = 00°00'08.8"	H.P. = 01°01'21.7"

External/Internal
Contacts of Penumbra

P1 = 08:15:56.1 UT1
P4 = 12:02:51.0 UT1

Circumstances at Greatest Eclipse
Lat. = 71°02.7'N	Sun Alt. = 0.0°
Long. = 101°24.7'W	Sun Azm. = 49.7°

ΔT = 98.9 s

www.EclipseWise.com

©2020 F. Espenak

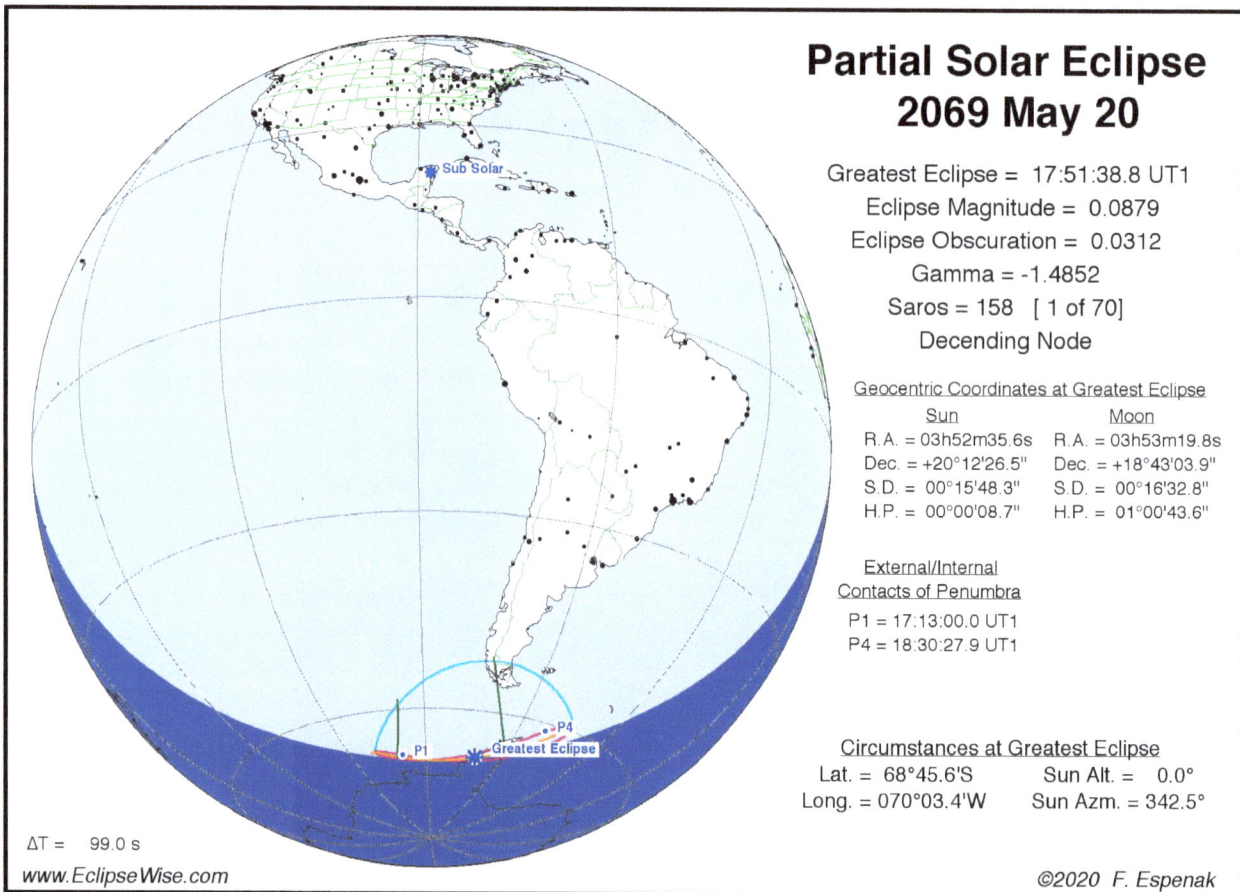

Partial Solar Eclipse
2069 May 20

Greatest Eclipse = 17:51:38.8 UT1
Eclipse Magnitude = 0.0879
Eclipse Obscuration = 0.0312
Gamma = -1.4852
Saros = 158 [1 of 70]
Decending Node

Geocentric Coordinates at Greatest Eclipse
Sun	Moon
R.A. = 03h52m35.6s	R.A. = 03h53m19.8s
Dec. = +20°12'26.5"	Dec. = +18°43'03.9"
S.D. = 00°15'48.3"	S.D. = 00°16'32.8"
H.P. = 00°00'08.7"	H.P. = 01°00'43.6"

External/Internal
Contacts of Penumbra

P1 = 17:13:00.0 UT1
P4 = 18:30:27.9 UT1

Circumstances at Greatest Eclipse
Lat. = 68°45.6'S	Sun Alt. = 0.0°
Long. = 070°03.4'W	Sun Azm. = 342.5°

ΔT = 99.0 s

www.EclipseWise.com

©2020 F. Espenak

Partial Solar Eclipse
2069 Oct 15

Greatest Eclipse = 04:18:17.0 UT1

Eclipse Magnitude = 0.5298

Eclipse Obscuration = 0.4130

Gamma = -1.2524

Saros = 125   [57 of 73]

Ascending Node

Geocentric Coordinates at Greatest Eclipse

	Sun	Moon
R.A. =	13h22m54.2s	13h21m37.6s
Dec. =	-08°43'06.9"	-09°48'03.1"
S.D. =	00°16'02.2"	00°14'45.3"
H.P. =	00°00'08.8"	00°54'09.1"

External/Internal Contacts of Penumbra

P1 = 02:26:17.8 UT1
P4 = 06:09:58.5 UT1

Circumstances at Greatest Eclipse

Lat. = 71°38.6'S	Sun Alt. = 0.0°
Long. = 005°36.4'W	Sun Azm. = 118.8°

ΔT =   99.3 s

www.EclipseWise.com

©2020 F. Espenak

Total Solar Eclipse
2070 Apr 11

Greatest Eclipse = 02:34:29.7 UT1

Eclipse Magnitude = 1.0472

Eclipse Obscuration = 1.0965

Gamma = 0.3652

Saros = 130   [55 of 73]

Decending Node

Geocentric Coordinates at Greatest Eclipse

	Sun	Moon
R.A. =	01h19m45.0s	01h19m20.0s
Dec. =	+08°24'18.3"	+08°45'25.7"
S.D. =	00°15'57.8"	00°16'27.4"
H.P. =	00°00'08.8"	01°00'23.9"

External/Internal Contacts of Penumbra

P1 = 23:58:06.3 UT1
P2 = 02:05:07.0 UT1
P3 = 03:03:35.5 UT1
P4 = 05:10:44.0 UT1

External/Internal Contacts of Umbra

U1 = 00:56:11.6 UT1
U2 = 00:57:57.9 UT1
U3 = 04:10:51.2 UT1
U4 = 04:12:41.2 UT1

Circumstances at Greatest Eclipse

Lat. = 29°03.3'N	Sun Alt. = 68.4°
Long. = 134°55.2'E	Sun Azm. = 161.7°
Path Width = 168.2 km	Duration = 04m04.2s

ΔT =   99.7 s

www.EclipseWise.com

©2020 F. Espenak

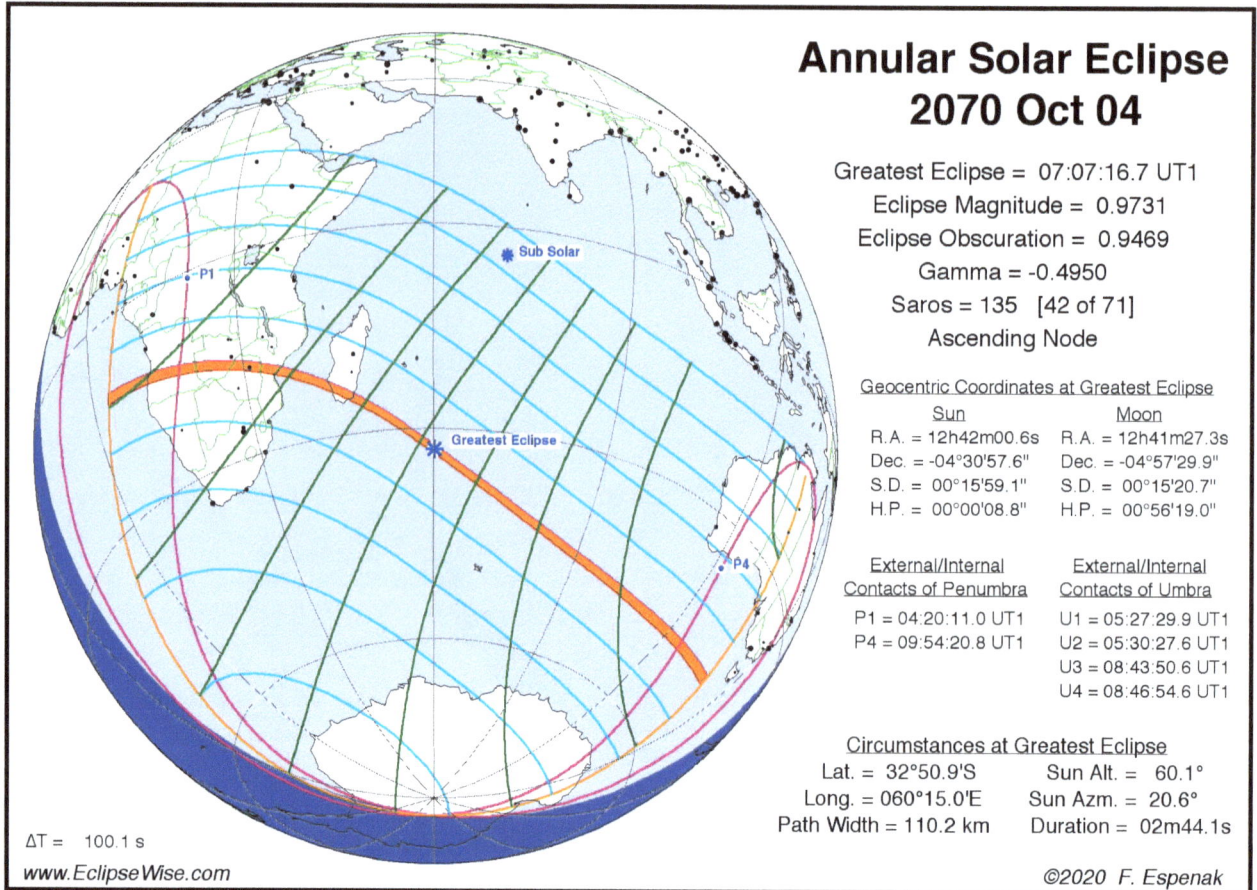

Annular Solar Eclipse
2070 Oct 04

Greatest Eclipse = 07:07:16.7 UT1
Eclipse Magnitude = 0.9731
Eclipse Obscuration = 0.9469
Gamma = -0.4950
Saros = 135   [42 of 71]
Ascending Node

Geocentric Coordinates at Greatest Eclipse

	Sun	Moon
R.A. =	12h42m00.6s	12h41m27.3s
Dec. =	-04°30'57.6"	-04°57'29.9"
S.D. =	00°15'59.1"	00°15'20.7"
H.P. =	00°00'08.8"	00°56'19.0"

External/Internal Contacts of Penumbra	External/Internal Contacts of Umbra
P1 = 04:20:11.0 UT1	U1 = 05:27:29.9 UT1
P4 = 09:54:20.8 UT1	U2 = 05:30:27.6 UT1
	U3 = 08:43:50.6 UT1
	U4 = 08:46:54.6 UT1

Circumstances at Greatest Eclipse

Lat. = 32°50.9'S	Sun Alt. = 60.1°
Long. = 060°15.0'E	Sun Azm. = 20.6°
Path Width = 110.2 km	Duration = 02m44.1s

ΔT =   100.1 s

Key to Catalog of Solar Eclipses

The following catalog lists all solar eclipses for a 25-year period.

A brief description of each parameter in the catalog appears below.

Date — Gregorian date of Greatest Eclipse

Greatest Eclipse — Universal Time of Greatest Eclipse

Saros — Saros Series Number of the eclipse

Type — Solar Eclipse Type

        P = Partial Solar Eclipse
        A = Annular Solar Eclipse
        T = Total Solar Eclipse
        H = Hybrid Solar Eclipse

        + = Non-central eclipse of with no northern limit
        – = Non-central eclipse of with no southern limit

        b = Saros series begins (first eclipse in a Saros series)
        e = Saros series ends (last eclipse in a Saros series)

Gamma — minimum distance from the axis of the lunar shadow to the center of Earth

Mag — Eclipse Magnitude; fraction of the Sun's diameter obscured by the Moon

Lat & Long — latitude and longitude where the Sun appears in the zenith at greatest eclipse

Alt — altitude of the Sun at greatest eclipse

Duration — Central Line Duration (minutes. seconds) at greatest eclipse

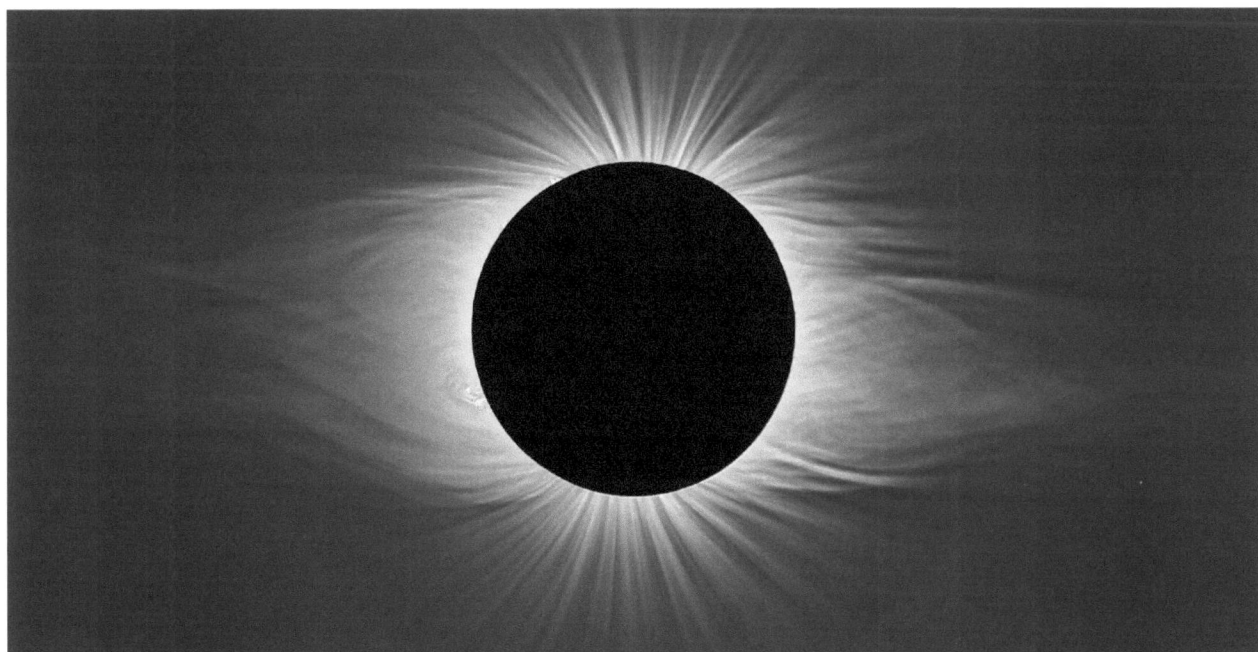

Photo 1–8 The solar corona is revealed during totality. Total Solar Eclipse of 2018 Jul 03. ©2018 F. Espenak

Catalog of Solar Eclipses: 2061 to 2085

Date	Greatest Eclipse	Saros	Type	Gamma	Mag	Lat	Long	Alt	Duration
2061 Apr 20	2:55:16	149	T	0.9578	1.0476	64.5N	59.1E	16	02m37s
2061 Oct 13	10:30:36	154	A	-0.9639	0.9469	62.1S	54.5W	15	03m41s
2062 Mar 11	4:24:43	121	P	-1.0238	0.9331	61.0S	147.2W	0	
2062 Sep 03	8:52:54	126	P	1.0192	0.9749	61.3N	150.2E	0	
2063 Feb 28	7:41:56	131	A	-0.3360	0.9293	25.2S	77.6E	70	07m41s
2063 Aug 24	1:20:36	136	T	0.2771	1.0750	25.6N	168.3E	74	05m49s
2064 Feb 17	6:58:48	141	A	0.3596	0.9262	7.0N	69.6E	69	08m56s
2064 Aug 12	17:44:31	146	T	-0.4652	1.0495	10.9S	96.1W	62	04m28s
2065 Feb 05	9:50:50	151	P	1.0336	0.9123	62.2N	22.0W	0	
2065 Jul 03	17:32:17	118	P	1.4619	0.1639	64.8N	71.7E	0	
2065 Aug 02	5:32:41	156	P	-1.2758	0.4903	62.7S	46.4E	0	
2065 Dec 27	8:38:19	123	P	-1.0688	0.8769	65.4S	149.3W	0	
2066 Jun 22	19:24:11	128	A	0.7330	0.9435	70.1N	96.5W	43	04m40s
2066 Dec 17	0:22:03	133	T	-0.4043	1.0416	47.4S	175.6E	66	03m14s
2067 Jun 11	20:40:49	138	A	-0.0387	0.9670	21.0N	130.3W	88	04m05s
2067 Dec 06	14:02:05	143	H	0.2845	1.0011	6.0S	32.5W	74	00m08s
2068 May 31	3:55:01	148	T	-0.7970	1.0110	31.0S	123.1E	37	01m06s
2068 Nov 24	21:30:51	153	P	1.0299	0.9109	68.5N	131.2W	0	
2069 Apr 21	10:09:30	120	P	1.0624	0.8992	71.0N	101.4W	0	
2069 May 20	17:51:39	158	Pb	-1.4852	0.0879	68.8S	70.1W	0	
2069 Oct 15	4:18:17	125	P	-1.2524	0.5298	71.6S	5.6W	0	
2070 Apr 11	2:34:30	130	T	0.3652	1.0472	29.1N	134.9E	68	04m04s
2070 Oct 04	7:07:17	135	A	-0.4950	0.9731	32.8S	60.3E	60	02m44s
2071 Mar 31	14:59:26	140	A	-0.3739	0.9919	16.7S	37.2W	68	00m52s
2071 Sep 23	17:18:47	145	T	0.2620	1.0333	14.2N	76.9W	75	03m11s
2072 Mar 19	20:08:50	150	P	-1.1405	0.7199	72.2S	30.5W	0	
2072 Sep 12	8:57:39	155	T	0.9655	1.0558	69.8N	101.8E	14	03m13s
2073 Feb 07	1:54:17	122	P	1.1651	0.6768	70.5N	114.7E	0	
2073 Aug 03	17:13:41	127	T	-0.8763	1.0294	43.2S	89.6W	28	02m29s
2074 Jan 27	6:42:33	132	A	0.4251	0.9798	6.6N	78.6E	65	02m21s
2074 Jul 24	3:08:49	137	A	-0.1242	0.9838	12.8N	133.6E	83	01m57s
2075 Jan 16	18:34:21	142	T	-0.2799	1.0311	37.2S	94.3W	74	02m42s
2075 Jul 13	6:04:00	147	A	0.6583	0.9467	63.1N	95.0E	49	04m45s
2076 Jan 06	10:05:43	152	T	-0.9373	1.0342	87.2S	173.9W	20	01m49s
2076 Jun 01	17:29:37	119	P	-1.3897	0.2897	64.4S	51.4W	0	
2076 Jul 01	6:48:59	157	P	1.4005	0.2746	67.0N	98.3W	0	
2076 Nov 26	11:41:16	124	P	1.1401	0.7315	63.7N	39.9E	0	
2077 May 22	2:44:20	129	T	-0.5725	1.0290	13.1S	148.1E	55	02m54s
2077 Nov 15	17:06:10	134	A	0.4705	0.9371	7.8N	71.0W	62	07m54s
2078 May 11	17:55:08	139	T	0.1838	1.0701	28.1N	93.9W	79	05m40s
2078 Nov 04	16:53:58	144	A	-0.2285	0.9255	27.8S	83.5W	77	08m29s
2079 May 01	10:48:26	149	T	0.9081	1.0512	66.2N	46.5W	24	02m55s
2079 Oct 24	18:09:34	154	A	-0.9243	0.9484	63.4S	160.8W	22	03m39s
2080 Mar 21	12:18:27	121	P	-1.0578	0.8734	60.9S	85.7E	0	
2080 Sep 13	16:36:21	126	P	1.0724	0.8743	61.1N	25.6E	0	
2081 Mar 10	15:21:42	131	A	-0.3653	0.9304	22.4S	36.9W	68	07m36s
2081 Sep 03	9:05:41	136	T	0.3379	1.0720	24.6N	53.4E	70	05m33s
2082 Feb 27	14:45:10	141	A	0.3361	0.9298	9.4N	47.3W	70	08m12s
2082 Aug 24	1:14:30	146	T	-0.4004	1.0452	10.3S	151.6E	66	04m01s
2083 Feb 16	18:04:46	151	P	1.0170	0.9433	61.6N	154.3W	0	
2083 Jul 15	0:12:32	118	Pe	1.5465	0.0169	64.0N	37.9W	0	
2083 Aug 13	12:32:50	156	P	-1.2064	0.6146	62.1S	67.7W	0	
2084 Jan 07	17:28:32	123	P	-1.0715	0.8723	64.4S	68.3E	0	
2084 Jul 03	1:48:34	128	A	0.8208	0.9421	75.0N	169.3W	35	04m25s
2084 Dec 27	9:11:56	133	T	-0.4094	1.0396	47.3S	47.5E	66	03m04s
2085 Jun 22	3:19:23	138	A	0.0453	0.9704	26.2N	131.0E	87	03m29s
2085 Dec 16	22:35:55	143	A	0.2786	0.9971	7.3S	161.0W	74	00m19s

Photo 2–1 Phases of the total lunar eclipse of 2014 Apr 15 are captured in this multiple exposure sequence. ©2014 F. Espenak

Section 2: Lunar Eclipses

Introduction

The Moon orbits Earth once every 29.5306 days with respect to the Sun. Over the course of its orbit, the Moon's changing position relative to the Sun results in its familiar phases: New Moon > First Quarter > Full Moon > Last Quarter > New Moon. The New Moon phase is the only one not visible because the illuminated side of the Moon points away from Earth.

During Full Moon, the Moon appears opposite the Sun in the sky. It rises as the Sun sets and is visible throughout the night. The Full Moon sets in the morning just as the Sun rises. This geometry occurs when the Moon is 180° from the Sun as seen from Earth. It corresponds to the direction Earth casts its shadow into space.

The Moon's orbit is tilted about 5.1° to Earth's orbit around the Sun. As seen from Earth, the points where the two orbits appear to cross are called the nodes. When the Full Moon occurs near one of these nodes, the Moon can pass through some portion of Earth's shadow and a lunar eclipse occurs.

Earth's shadow has two cone-shaped components, one nested inside the other. The outer or penumbral shadow is a zone where the Sun's rays are partially blocked., The inner or umbral shadow is a region where direct rays from the Sun are completely blocked.

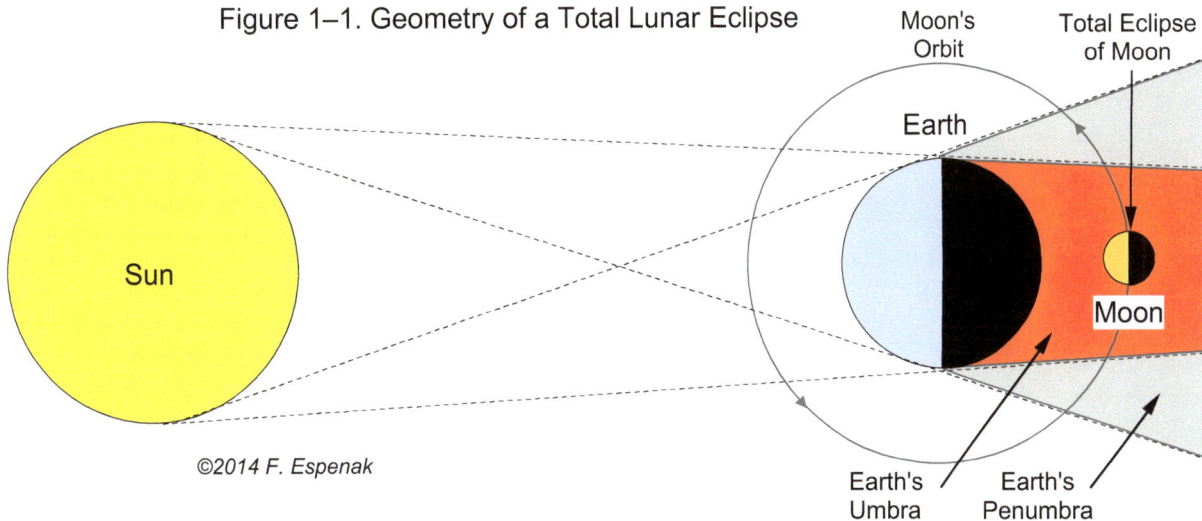

Figure 1–1. Geometry of a Total Lunar Eclipse

©2014 F. Espenak

Figure 2–1 illustrates the geometry of a total lunar eclipse. A partial eclipse is visible if only part of the Moon enters Earth's umbral shadow. If the Moon passes through the penumbral shadow but misses the umbral shadow, then a penumbral eclipse occurs.

Photo 2–2 Examples of the visual appearance of a penumbral, partial, and total lunar eclipse. ©2020 F. Espenak

Types of Lunar Eclipses

There are three types of lunar eclipses:

1. **Penumbral Lunar Eclipse** — The Moon passes through Earth's faint penumbral shadow. Penumbral eclipses are of minor interest since they are quite subtle and difficult to observe.
2. **Partial Lunar Eclipse** — A portion of the Moon passes through Earth's dark umbral shadow. The remaining part of the Moon appears bright even though it lies deep within the penumbra. Partial eclipses are easy to see, even with the unaided eye.
3. **Total Lunar Eclipse** — The entire Moon passes through Earth's umbral shadow. Total eclipses are quite striking for the vibrant range of color of the Moon during the total phase, referred to as totality.

Figure 1–2. Types of Lunar Eclipses

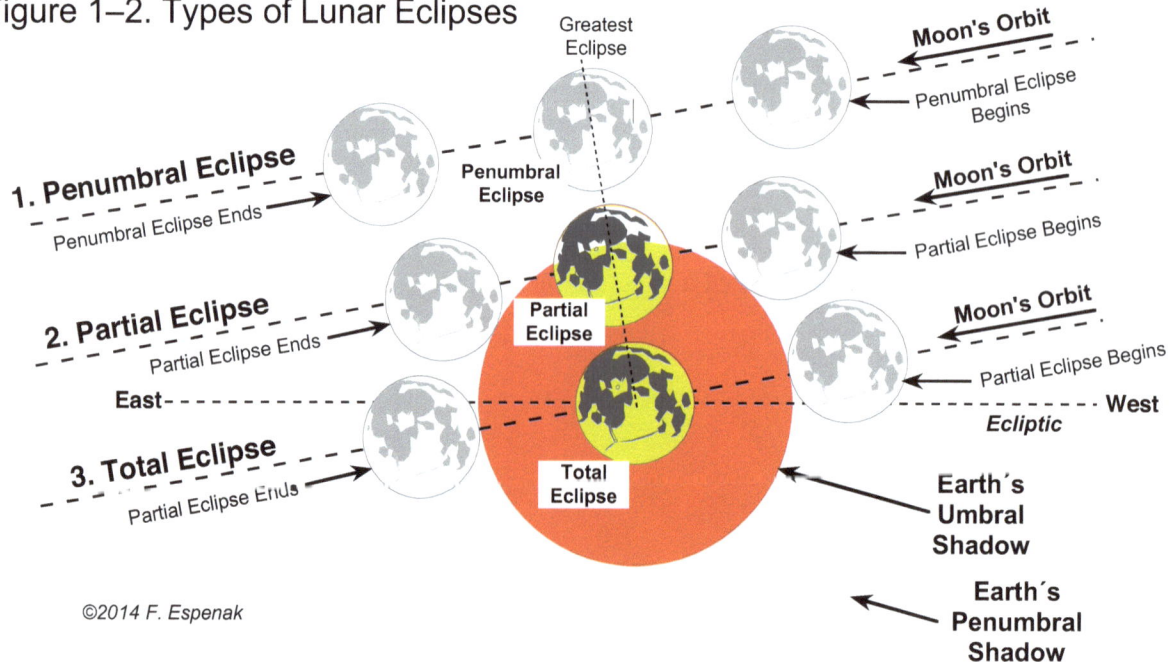

Figure 2–2 illustrates the three types of lunar eclipses as seen from Earth. 1) A penumbral eclipse occurs when the Moon passes through the penumbra but completely misses the umbra. 2) A partial eclipse occurs if some portion of the Moon enters the umbra. 3) A total lunar eclipse takes place when the entire disk of the Moon enters the umbral shadow.

Photo 2–3 Penumbral lunar eclipse of 2002 Nov 20. Left: Moon before the eclipse begins. Right: mid-eclipse when 88.9% of the Moon's is in the penumbral shadow. ©2002 F. Espenak

Visual Appearance of Penumbral Lunar Eclipses

The visual appearance of penumbral, partial and total lunar eclipses differs significantly. While penumbral eclipses are pale and difficult to see, partial eclipses are easy naked-eye events while total eclipses are colorful and dramatic.

Earth's penumbral shadow forms a diverging cone that expands into space away from the Sun. Within this zone, Earth blocks part but not all of the Sun's disk. Some portion of the Sun's direct rays continue to reach the Moon during a penumbral eclipse.

The early and late stages of a penumbral eclipse are completely invisible to the eye. It is only when about 2/3 of the Moon's disk has entered the penumbral shadow that a skilled observer can detect a faint shading across the Moon.

Even when 90% of the Moon is immersed in the penumbra, approximately 10% of the Sun's rays still reach the Moon's deepest limb. Under such conditions, the Moon remains relatively bright with only a subtle shadow gradient across its disk.

Photo 2–4 Time sequence of the partial lunar eclipse of 2012 Jun 04. ©2012 F. Espenak

Visual Appearance of Partial Lunar Eclipses

Compared to penumbral eclipses, partial eclipses are easy to see with the naked eye. The lunar limb extending into the umbral shadow appears very dark or even black. This is due to a contrast effect since the remaining portion of the Moon in the penumbra is hundreds of times brighter. Because the umbral shadow's diameter is about 2.7 times the Moon's diameter, it appears as though a semi-circular bite has been taken out of the Moon.

Aristotle (384–322 BCE) first proved that Earth was round using the curved umbral shadow seen at partial eclipses. In comparing observations of several eclipses, he noted that Earth's shadow was round no matter where the eclipse took place, whether the Moon was high in the sky or low near the horizon. Aristotle reasoned that only a sphere casts a round shadow from every angle.

*Photo 2–5 The beginning, middle and end of totality during the total lunar eclipse of
2004 October 28. ©2004 F. Espenak*

Visual Appearance of Total Lunar Eclipses

A total lunar eclipse is the most dramatic and visually compelling type of lunar eclipse. The Moon's appearance can vary enormously throughout the period of totality and from one eclipse to the next. The geometry of the Moon's path through the umbra plays a significant role in determining the appearance of totality. The effect that Earth's atmosphere has on a total eclipse is not as apparent. Although the physical mass of Earth blocks all direct sunlight from the umbra, the planet's atmosphere filters, attenuates and bends some of the Sun's rays into the shadow.

The molecules in Earth's atmosphere scatter short wavelength light (i.e., yellow, green, blue) more than long wavelength light (i.e., orange, red). The same process responsible for making sunsets red also gives total lunar eclipses their characteristic ruddy color. The exact appearance can vary widely in both hue and brightness.

Because the lowest layers of the atmosphere are the thickest, they absorb more sunlight and refract it through larger angles. About 75% of the atmosphere's mass is concentrated in the bottom 10 kilometers (troposphere) as well as most of the water vapor, which can form massive clouds that block even more light. Just above the troposphere lies the stratosphere (10 to 50 kilometers), a rarified zone above most of the planet's weather systems. The stratosphere is subject to important photochemical reactions due to the high level of solar ultraviolet radiation that penetrates the region. The troposphere and stratosphere act together as a ring-shaped lens that refracts heavily reddened sunlight into Earth's umbral shadow. Since the higher stratospheric layers contain less gas, they refract sunlight through progressively smaller angles into the outer parts of the umbra. In contrast, denser tropospheric layers refract sunlight through larger angles to reach the inner parts of the umbra.

Because of this lensing effect, the amount of light refracted into the umbra tends to decrease radially from the edge to the center. Inhomogeneities from varying amounts of cloud and dust at differing latitudes can cause significant variations in brightness throughout the umbra.

Besides water (cloud, mist, precipitation), Earth's atmosphere also contains aerosols or tiny particles of organic debris, meteoric dust, volcanic ash and photochemical droplets. This material attenuates sunlight before it is refracted into the umbra. For instance, major volcanic eruptions in 1963 (Agung) and 1982 (El Chichon) each dumped large quantities of gas and ash into the stratosphere and were followed by several years of dark eclipses.

The 1991 eruption of Pinatubo in the Philippines had a similar effect. While most of the solid ash fell to Earth several days after circulating through the troposphere, a sizable volume of sulfur dioxide (SO_2) reached the stratosphere where it interacts with water vapor and eventually produces sulfuric acid (H_2SO_4). This high-altitude volcanic haze layer severely attenuates sunlight that must travel several hundred kilometers horizontally through the layer before being refracted into the umbral shadow. Thus, total eclipses following large volcanic eruptions are particularly dark. The total lunar eclipse of 1992 Dec 09 (1½ years after Pinatubo) was so dark that it was difficult to see the Moon's dull gray disk with the naked eye.

All total eclipses begin with penumbral and partial phases. After the total phase, the eclipse ends with more partial and penumbral phases. Lunar eclipses are completely safe to view and require none of the precautions needed for viewing solar eclipses (like special filters). The best views of a lunar eclipse are with binoculars and the naked eye.

Figure 2–1. Lunar Eclipse Contacts

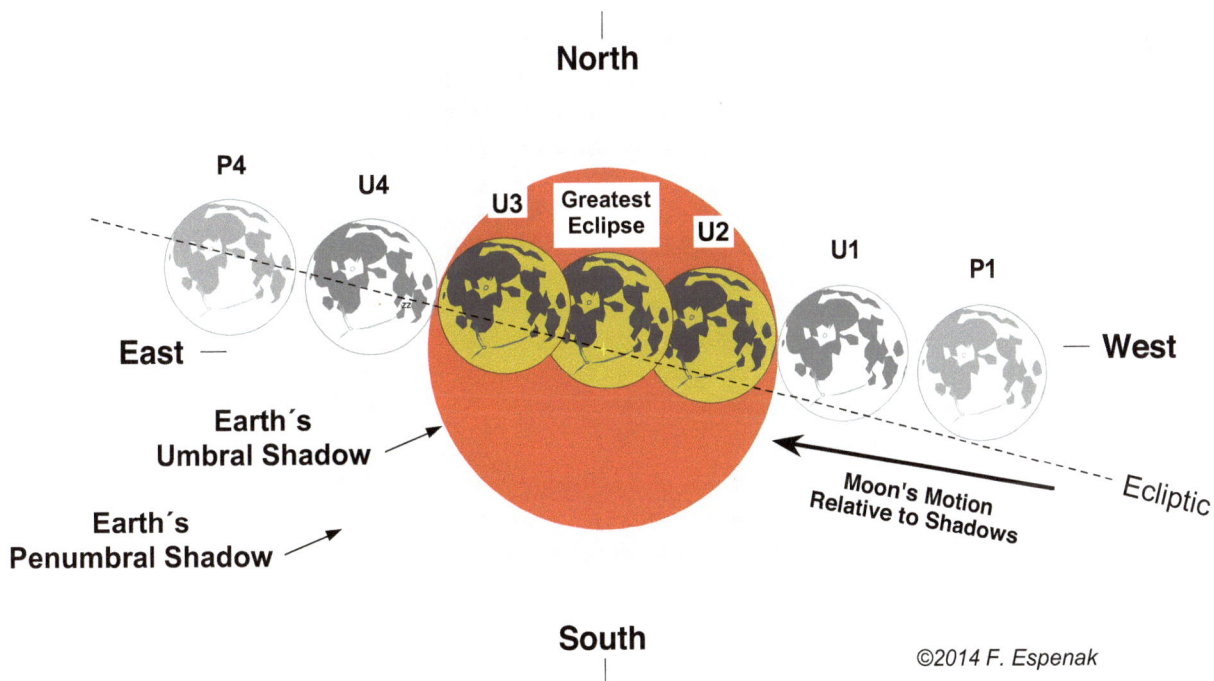

©2014 F. Espenak

Figure 2–3 illustrates the six contacts for a total lunar eclipse. These correspond to the instants when the Moon's disk is externally tangent to the penumbra (P1 and P4), or either externally or internally tangent to the umbra (U1, U2, U3, and U4). Partial eclipses do not have contacts U2 and U3, while penumbral eclipses only have contacts P1 and P4.

Lunar Eclipse Contacts

During the course of a lunar eclipse, the instants when the Moon's disk becomes tangent to Earth's shadows are known as eclipse contacts. They mark the primary stages or phases of a lunar eclipse (see figure 2-3) and are defined as follows.

P1 — Penumbral Eclipse Begins   (Instant of first exterior tangency of the Moon with the Penumbra)
U1 — Partial Eclipse Begins   (Instant of first exterior tangency of the Moon with the Umbra)

31

U2 — Total Eclipse Begins   (Instant of first interior tangency of the Moon with the Umbra)
U3 — Total Eclipse Ends   (Instant of last interior tangency of the Moon with the Umbra)
U4 — Partial Eclipse Ends   (Instant of last exterior tangency of the Moon with the Umbra)
P4 — Penumbral Eclipse Ends   (Instant of last exterior tangency of the Moon with the Penumbra)

Penumbral eclipses only have first and last contacts (i.e., P1 and P4) although neither of these events is observable.

In addition to P1 and P4, partial eclipses also have contacts U1 and U4 when the partial phases begin and end.

Total lunar eclipses have all six contacts. Contacts U2 and U3 mark the instants when the Moon's entire disk is first and last internally tangent to the umbra. These are the times when the total phase of the eclipse begins and ends..

The instant of greatest eclipse occurs when the Moon passes closest to the shadow axis. This is the maximum phase of the eclipse when the Moon is at its deepest position within either the penumbral or umbral shadow.

Photo 2–6 Five minutes before the start of totality (2018 Jan 31), the Moon is bathed in an orange-red light.
The narrow rim outside the umbra and still in sunlight appears brilliant white. ©2018 F. Espenak

Enlargement of Earth's Shadows

In 1707, Philippe de La Hire made a curious observation about Earth's umbra. The predicted radius of the shadow needed to be enlarged by about 1/41 in order to fit timings made during a lunar eclipse. The enlargement is attributed to Earth's atmosphere, which becomes increasingly less transparent at lower levels. Additional observations over the next two centuries revealed that the shadow enlargement was somewhat variable from one eclipse to the next.

William Chauvenet (1891) formulated one method to account for the shadow enlargement while André-Louis Danjon (1951) devised another. Both methods assumed a circular cross section for the umbral shadow. However, Earth is flattened at the poles and bulges at the Equator, so an oblate spheroid more closely represents its shape. The projection of each of the planet's shadows is an ellipse rather than a circle. Furthermore, Earth's axial tilt towards or away from the Sun throughout the year means the elliptical shape of the penumbral and umbral shadows varies as well.

In an analysis of 22,539 observations made at 94 lunar eclipses from 1842 to 2011, Herald and Sinnott[7] (2014) found that the size and shape of the umbra are consistent with an oblate spheroid at the time of each eclipse, enlarged by the empirically determined occulting layer that uniformly surrounds Earth. The effective height of this layer was found to be 87 kilometers. Based on this work, the authors developed a new method to calculate the shadow enlargement including their elliptical shape.

This new method is the most rigorous and accurate procedure to date. The *Eclipse Almanac* uses it in the lunar eclipse predictions presented here.

Explanation of Lunar Eclipse Figures

There are 22 eclipses of the Moon during the period 2061 to 2070. A figure for each eclipse is included.

Each figure consists of two diagrams. The first one depicts the Moon's path through Earth's penumbral and umbral shadows (with Celestial North up). The second is a map showing the geographic visibility of each eclipse phase. All features in these diagrams are identified in the key on the next page.

The Moon's orbital motion with respect to the shadows is from west to east (right to left). Each phase of the eclipse is defined by the instant when the Moon's limb is externally or internally tangent to the penumbra or umbra. The six primary contacts of the Moon with the penumbral and umbral shadows are defined as follows.

P1 — Penumbral Eclipse Begins   (Instant of first exterior tangency of the Moon with the Penumbra)
U1 — Partial Eclipse Begins   (Instant of first exterior tangency of the Moon with the Umbra)
U2 — Total Eclipse Begins   (Instant of first interior tangency of the Moon with the Umbra)
U3 — Total Eclipse Ends   (Instant of last interior tangency of the Moon with the Umbra)
U4 — Partial Eclipse Ends   (Instant of last exterior tangency of the Moon with the Umbra)
P4 — Penumbral Eclipse Ends   (Instant of last exterior tangency of the Moon with the Penumbra)

Penumbral lunar eclipses have two primary contacts: P1 and P4, but neither is observable because the edge of the penumbra is indistinct and extremely faint.

In addition to the penumbral contacts, partial lunar eclipses have two more contacts as the Moon's limb enters and exits the umbral shadow: U1 and U4, respectively. These two contacts mark the instants when the partial phase of the eclipse begins and ends.

Total lunar eclipses undergo all six contacts. The two additional umbral contacts are the instants when the Moon's entire disk is first and last internally tangent to the umbra: U2 and U3, respectively. They mark the times when the total phase of the eclipse begins and ends.

The Moon passes closest to the shadow axis at the instant of greatest eclipse. This corresponds to the maximum phase of the eclipse and the Moon's position at this instant is also depicted in the path diagrams.

The equidistant cylindrical projection map shows the geographic region of visibility at each phase of the eclipse. This is accomplished using a series of curves showing where Moonrise and Moonset occur at each eclipse contact. The map is shaded to indicate eclipse visibility. The entire eclipse is visible from the zone with no shading. Conversely, none of the eclipse can be seen from the zone with the darkest shading.

At greatest eclipse, the Moon is deepest in Earth's shadow. The geographic location where the Moon appears in the zenith at greatest eclipse is shown by an asterisk.

---

[7] Herald, D., and Sinnott, R. W., "Analysis of Lunar Crater Timings, 1842–2011," *J. Br. Astron. Assoc.*, **124**, 5 (2014)

Parameters relevant to the eclipse appear in on the right side of each figure. The instant of greatest eclipse is the time (Universal Time[8] or UT1) when the Moon passes closest to the shadow axis. The penumbral and umbral magnitudes are the fractions of the Moon's diameter immersed in each shadow. Gamma is the minimum distance of the Moon's center from the axis of Earth's shadow at greatest eclipse. The Saros series of the eclipse, and the node of the Moon's orbit are given. Each contact time of the Moon's edge with the penumbral and umbral shadows is listed in Universal Time (UT1). Depending on the eclipse type, the duration of the penumbral, partial or total phases are given.

2–7 A time sequence shows the partial phases and totality during the total lunar eclipse of 2000 July 16 from Maui, Hawaii (Copyright ©2000 by Fred Espenak).

[8] Universal Time (UT1) is the modern-day replacement for Greenwich Mean Time and is based on Earth's rotation with respect to distant quasars.

Lunar Eclipse Figures

Key to Lunar Eclipse Figures

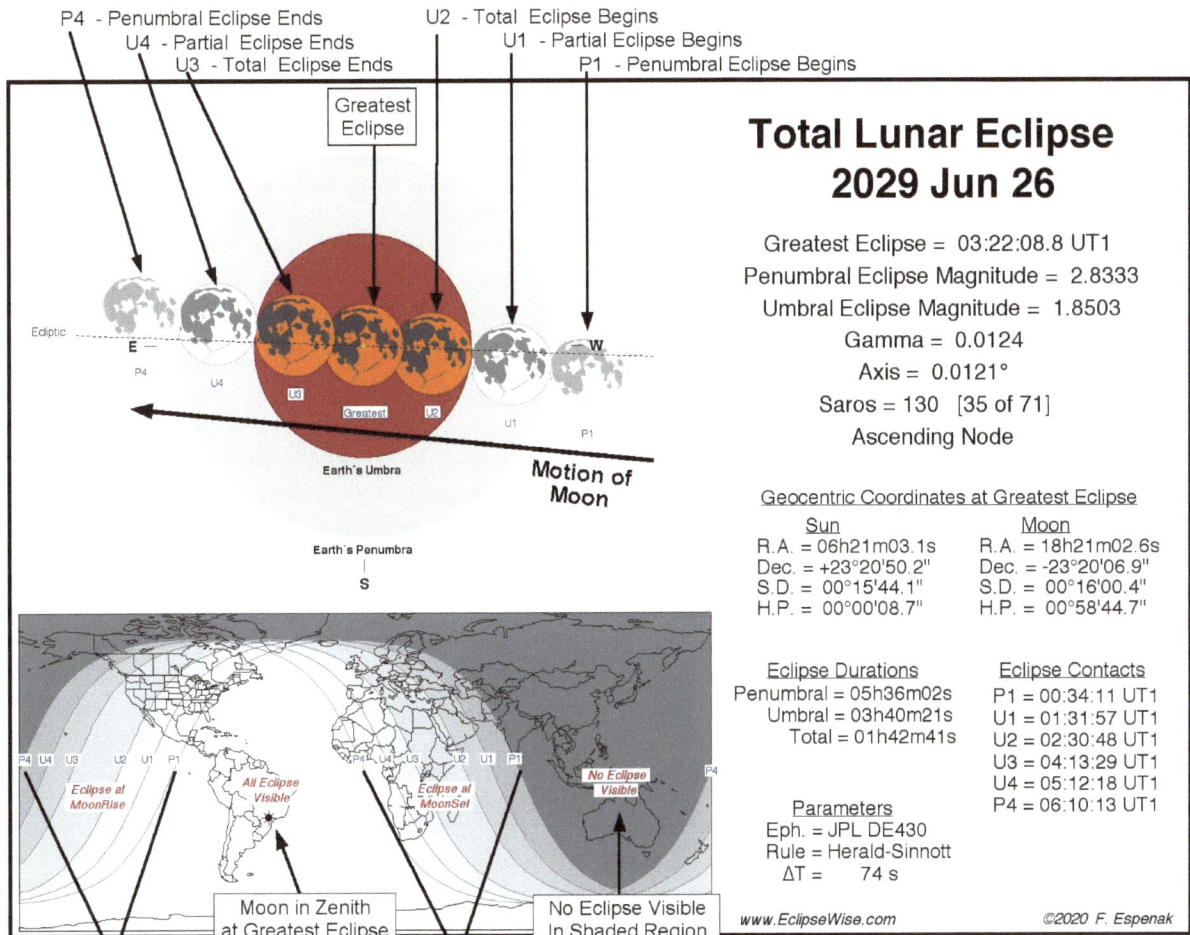

Total Lunar Eclipse
2029 Jun 26

Greatest Eclipse = 03:22:08.8 UT1
Penumbral Eclipse Magnitude = 2.8333
Umbral Eclipse Magnitude = 1.8503
Gamma = 0.0124
Axis = 0.0121°
Saros = 130 [35 of 71]
Ascending Node

Geocentric Coordinates at Greatest Eclipse

Sun	Moon
R.A. = 06h21m03.1s	R.A. = 18h21m02.6s
Dec. = +23°20'50.2"	Dec. = -23°20'06.9"
S.D. = 00°15'44.1"	S.D. = 00°16'00.4"
H.P. = 00°00'08.7"	H.P. = 00°58'44.7"

Eclipse Durations
Penumbral = 05h36m02s
Umbral = 03h40m21s
Total = 01h42m41s

Eclipse Contacts
P1 = 00:34:11 UT1
U1 = 01:31:57 UT1
U2 = 02:30:48 UT1
U3 = 04:13:29 UT1
U4 = 05:12:18 UT1
P4 = 06:10:13 UT1

Parameters
Eph. = JPL DE430
Rule = Herald-Sinnott
ΔT = 74 s

www.EclipseWise.com   ©2020 F. Espenak

Eclipse During Moon Rise
P1 - Penumbral Eclipse Begins
U1 - Partial Eclipse Begins
U2 - Total Eclipse Begins
U3 - Total Eclipse Ends
U4 - Partial Eclipse Ends
P4 - Penumbral Eclipse Ends

Eclipse During Moon Set
P1 - Penumbral Eclipse Begins
U1 - Partial Eclipse Begins
U2 - Total Eclipse Begins
U3 - Total Eclipse Ends
U4 - Partial Eclipse Ends
P4 - Penumbral Eclipse Ends

Layout of Lunar Eclipse Figure
Upper Left: Moon's Path Thru Earth's Shadows
Lower Left: Map of Eclipse Visibility
Right Side: Parameters for Lunar Eclipse

Explanation of Parameters Used in Lunar Eclipse Figures

Greatest Eclipse – The instant when the Moon passes closest to the axis of Earth's shadow cone (Universal Time[9])
Penumbral Eclipse Magnitude – Fraction of the Moon's diameter immersed in the penumbra at greatest eclipse.
Umbral Eclipse Magnitude – The fraction of the Moon's diameter immersed in the umbra at greatest eclipse.
Gamma – Minimum distance from the Moon's center to Earth's shadow axis (units of Earth's equatorial radius).
Axis – Minimum distance from the Moon's center to Earth's shadow axis (units of degrees).
Saros Series – The Saros series that the eclipse belongs to. The numbers in "[ ]" are the eclipse's sequential position and the number of eclipses in the Saros series.
Node – The orbital node near which the eclipse takes place (Ascending Node or Descending Node).
Geocentric Coordinates of the Sun and the Moon at Greatest Eclipse
  R.A. – Right Ascension   **S.D.** – Semi-Diameter (i.e. - radius)
  Dec. – Declination   **H.P.** – Horizontal Parallax
Eclipse Durations – Durations of Penumbral, Partial, and Total Eclipse.
Eclipse Contacts – Contact Times (Universal Time or UT1) of the Moon with the Penumbra and the Umbra
  P1, P4 – Start and End of the Penumbral Eclipse
  U1, U4 – Start and End of the Partial Eclipse
  U2, U3 – Start and End of the Total Eclipse

---

[9] Universal Time or UT1 is the modern replacement for Greenwich Mean Time

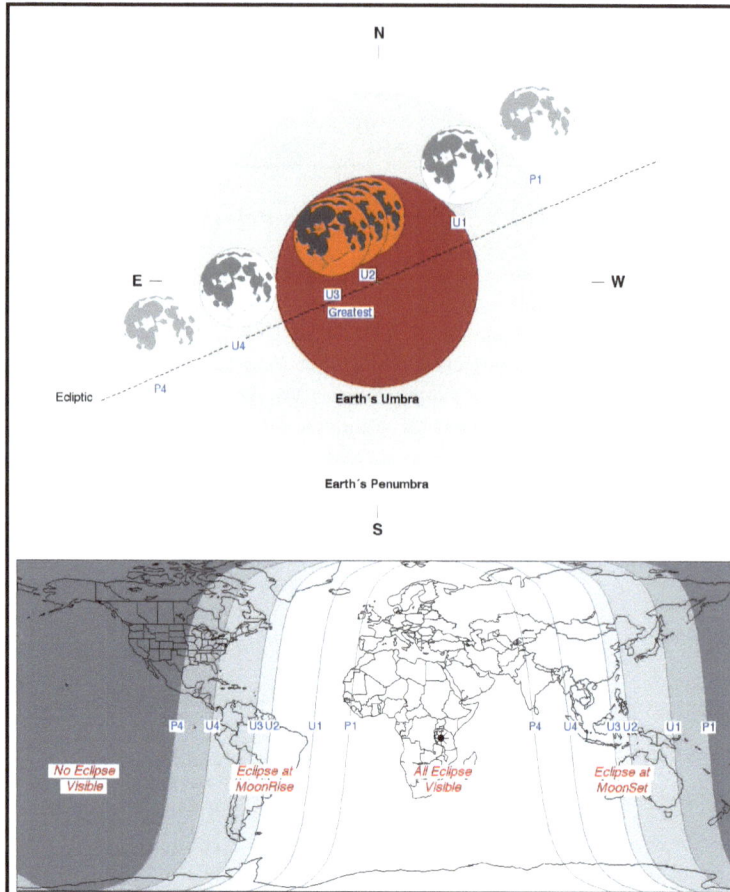

Total Lunar Eclipse
2061 Apr 04

Greatest Eclipse = 21:52:32.0 UT1
Penumbral Eclipse Magnitude = 2.1111
Umbral Eclipse Magnitude = 1.0407
Gamma = 0.4300
Axis = 0.3929°
Saros = 123 [55 of 72]
Decending Node

Geocentric Coordinates at Greatest Eclipse

Sun	Moon
R.A. = 00h57m38.4s	R.A. = 12h58m23.4s
Dec. = +06°09'23.4"	Dec. = -05°48'38.8"
S.D. = 00°15'59.4"	S.D. = 00°14'56.4"
H.P. = 00°00'08.8"	H.P. = 00°54'49.7"

Eclipse Durations	Eclipse Contacts
Penumbral = 05h56m01s	P1 = 18:54:33 UT1
Umbral = 03h30m36s	U1 = 20:07:13 UT1
Total = 00h32m45s	U2 = 21:36:08 UT1
	U3 = 22:08:54 UT1
	U4 = 23:37:48 UT1
Parameters	P4 = 00:50:34 UT1
Eph. = JPL DE430	
Rule = Herald-Sinnott	
ΔT = 93 s	

www.EclipseWise.com   ©2020 F. Espenak

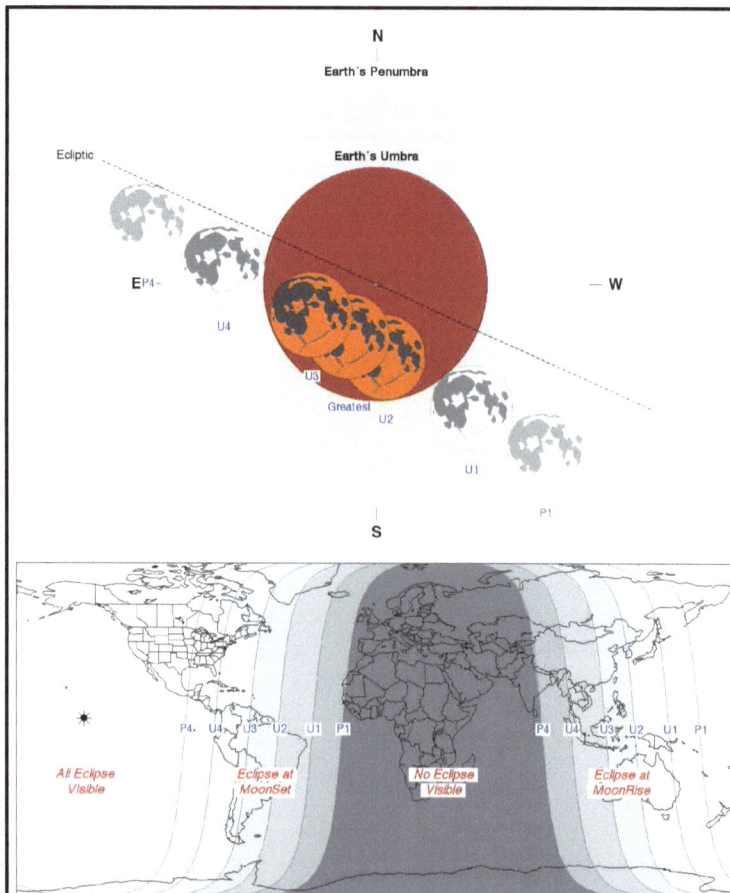

Total Lunar Eclipse
2061 Sep 29

Greatest Eclipse = 09:36:40.0 UT1
Penumbral Eclipse Magnitude = 2.1623
Umbral Eclipse Magnitude = 1.1687
Gamma = -0.3810
Axis = 0.3745°
Saros = 128 [43 of 71]
Ascending Node

Geocentric Coordinates at Greatest Eclipse

Sun	Moon
R.A. = 12h24m54.1s	R.A. = 00h25m37.7s
Dec. = -02°41'28.5"	Dec. = +02°21'48.8"
S.D. = 00°15'57.9"	S.D. = 00°16'04.1"
H.P. = 00°00'08.8"	H.P. = 00°58'58.5"

Eclipse Durations	Eclipse Contacts
Penumbral = 05h25m41s	P1 = 06:53:47 UT1
Umbral = 03h23m18s	U1 = 07:55:03 UT1
Total = 01h00m21s	U2 = 09:06:31 UT1
	U3 = 10:06:52 UT1
	U4 = 11:18:20 UT1
Parameters	P4 = 12:19:29 UT1
Eph. = JPL DE430	
Rule = Herald-Sinnott	
ΔT = 93 s	

www.EclipseWise.com   ©2020 F. Espenak

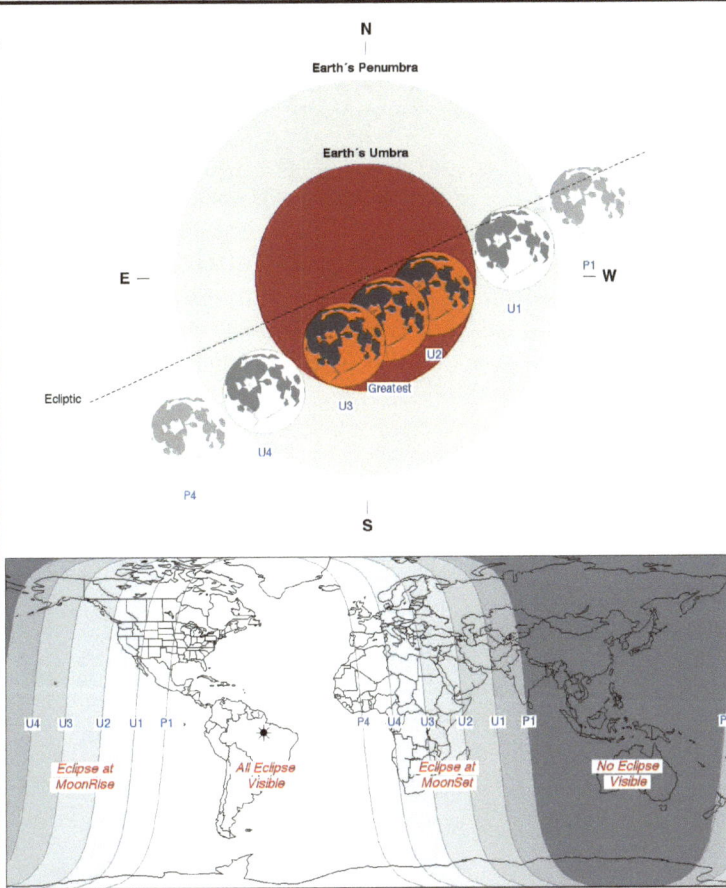

Total Lunar Eclipse
2062 Mar 25

Greatest Eclipse = 03:32:16.9 UT1
Penumbral Eclipse Magnitude = 2.2972
Umbral Eclipse Magnitude = 1.2762
Gamma = -0.3150
Axis = 0.3027°
Saros = 133 [29 of 71]
Decending Node

Geocentric Coordinates at Greatest Eclipse

Sun	Moon
R.A. = 00h17m31.2s	R.A. = 12h16m55.9s
Dec. = +01°53'45.5"	Dec. = -02°09'38.2"
S.D. = 00°16'02.5"	S.D. = 00°15'42.7"
H.P. = 00°00'08.8"	H.P. = 00°57'39.7"

Eclipse Durations	Eclipse Contacts
Penumbral = 05h39m15s	P1 = 00:42:42 UT1
Umbral = 03h32m12s	U1 = 01:46:09 UT1
Total = 01h15m45s	U2 = 02:54:23 UT1
	U3 = 04:10:08 UT1
	U4 = 05:18:21 UT1
Parameters	P4 = 06:21:57 UT1
Eph. = JPL DE430	
Rule = Herald-Sinnott	
ΔT = 94 s	

www.EclipseWise.com   ©2020 F. Espenak

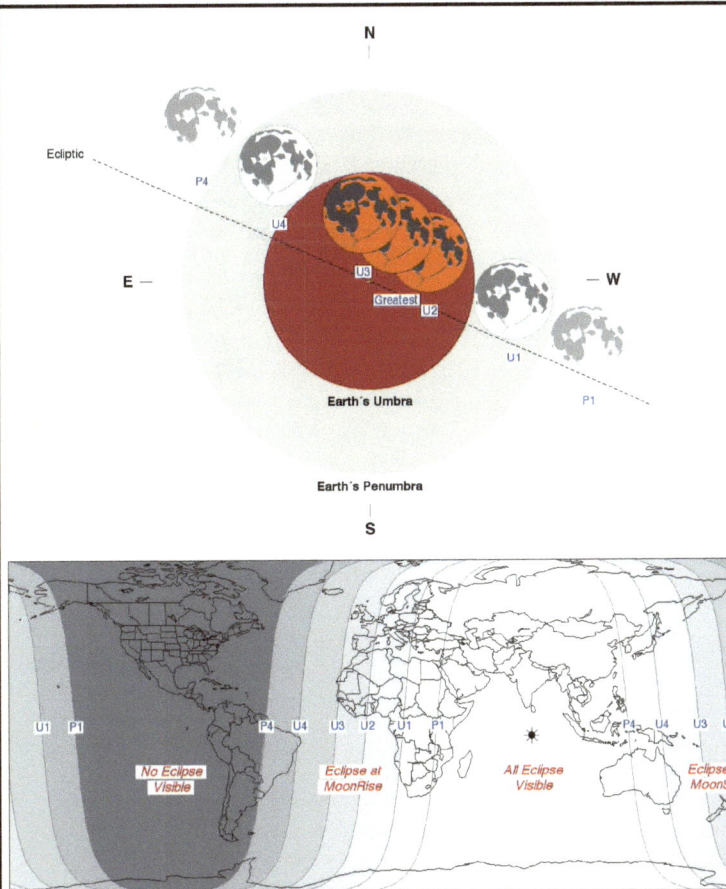

Total Lunar Eclipse
2062 Sep 18

Greatest Eclipse = 18:32:28.2 UT1
Penumbral Eclipse Magnitude = 2.2026
Umbral Eclipse Magnitude = 1.1563
Gamma = 0.3736
Axis = 0.3476°
Saros = 138 [31 of 82]
Ascending Node

Geocentric Coordinates at Greatest Eclipse

Sun	Moon
R.A. = 11h45m50.3s	R.A. = 23h45m09.8s
Dec. = +01°31'59.3"	Dec. = -01°13'45.9"
S.D. = 00°15'55.0"	S.D. = 00°15'12.7"
H.P. = 00°00'08.8"	H.P. = 00°55'49.7"

Eclipse Durations	Eclipse Contacts
Penumbral = 05h50m12s	P1 = 15:37:20 UT1
Umbral = 03h33m22s	U1 = 16:45:49 UT1
Total = 01h00m58s	U2 = 18:02:00 UT1
	U3 = 19:02:59 UT1
	U4 = 20:19:11 UT1
Parameters	P4 = 21:27:32 UT1
Eph. = JPL DE430	
Rule = Herald-Sinnott	
ΔT = 94 s	

www.EclipseWise.com   ©2020 F. Espenak

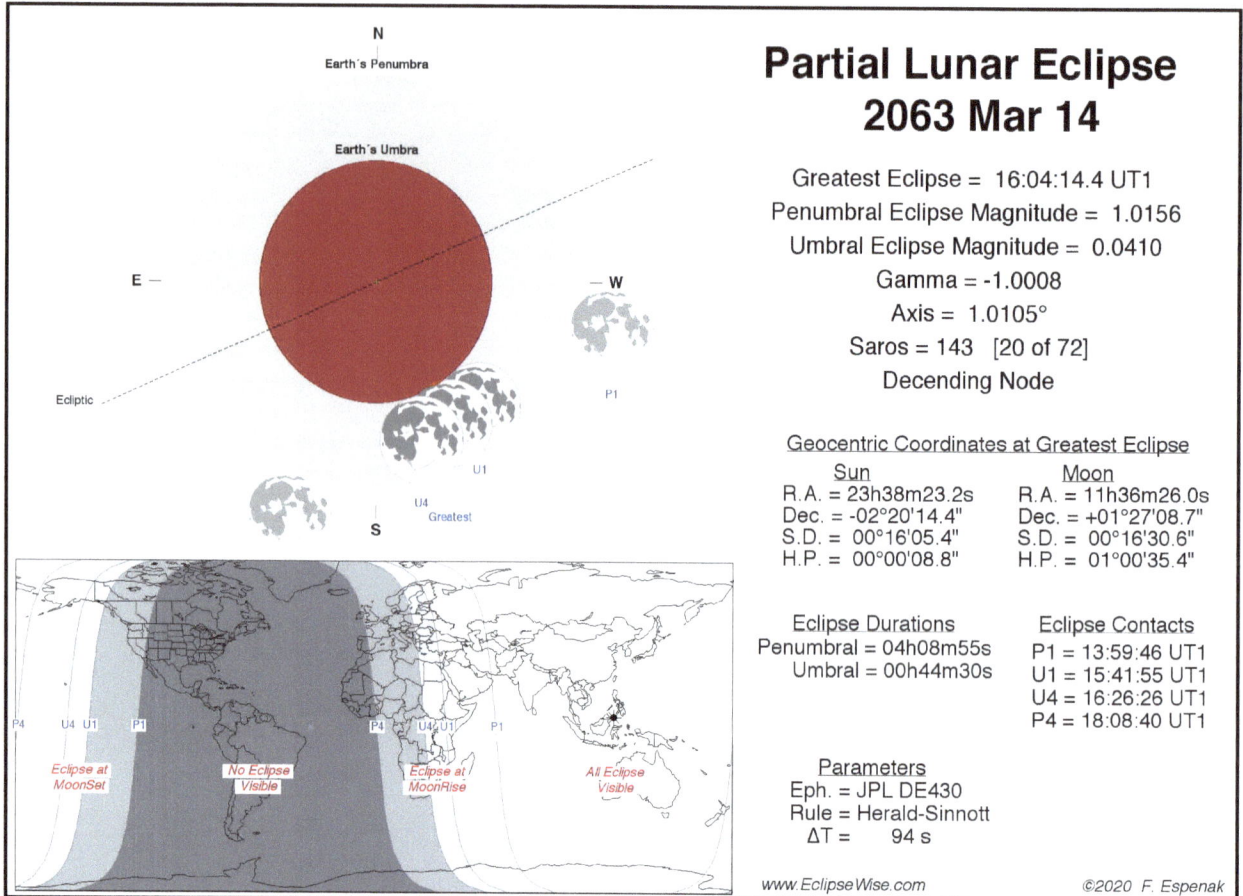

Partial Lunar Eclipse
2063 Mar 14

Greatest Eclipse = 16:04:14.4 UT1
Penumbral Eclipse Magnitude = 1.0156
Umbral Eclipse Magnitude = 0.0410
Gamma = -1.0008
Axis = 1.0105°
Saros = 143 [20 of 72]
Decending Node

Geocentric Coordinates at Greatest Eclipse

Sun	Moon
R.A. = 23h38m23.2s	R.A. = 11h36m26.0s
Dec. = -02°20'14.4"	Dec. = +01°27'08.7"
S.D. = 00°16'05.4"	S.D. = 00°16'30.6"
H.P. = 00°00'08.8"	H.P. = 01°00'35.4"

Eclipse Durations	Eclipse Contacts
Penumbral = 04h08m55s	P1 = 13:59:46 UT1
Umbral = 00h44m30s	U1 = 15:41:55 UT1
	U4 = 16:26:26 UT1
	P4 = 18:08:40 UT1

Parameters
Eph. = JPL DE430
Rule = Herald-Sinnott
ΔT = 94 s

www.EclipseWise.com   ©2020 F. Espenak

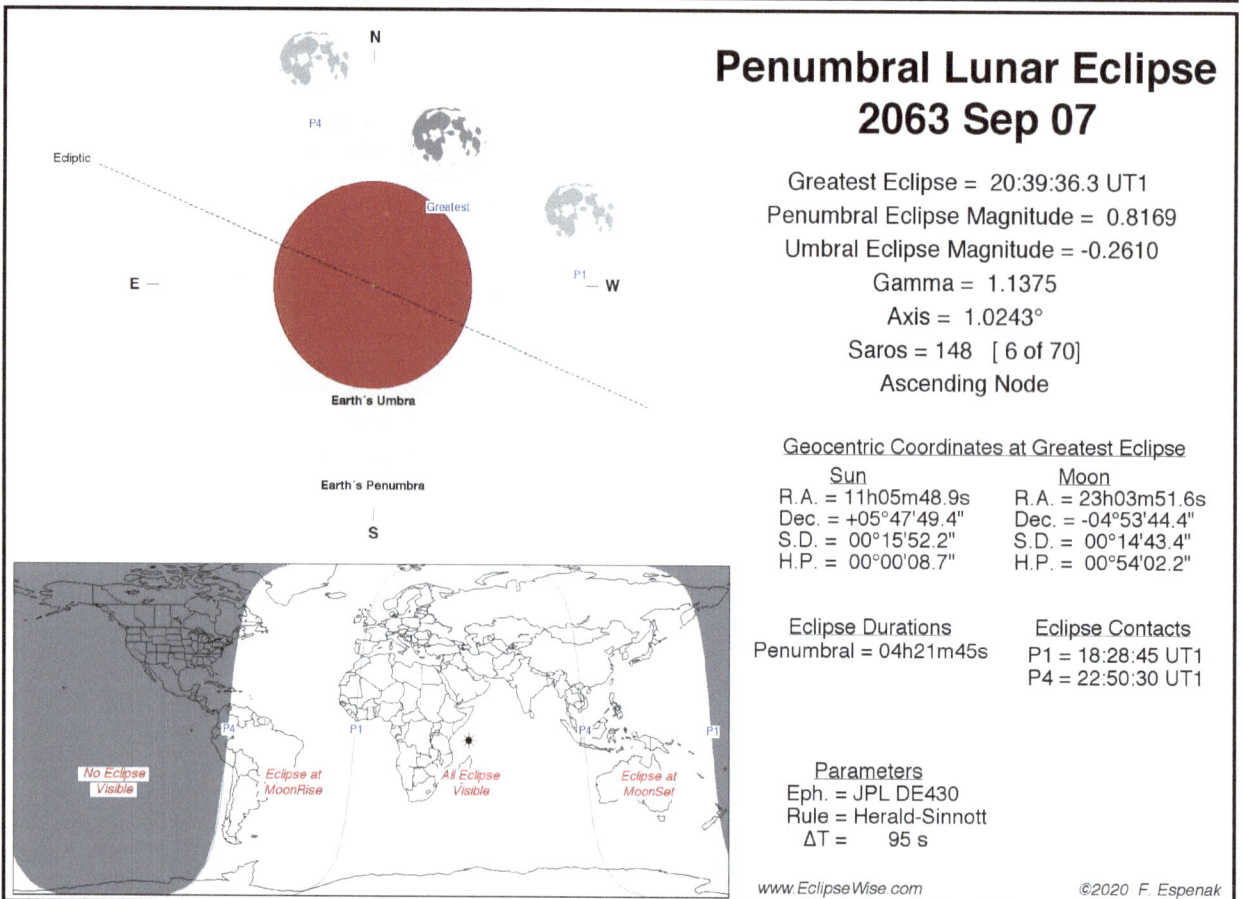

Penumbral Lunar Eclipse
2063 Sep 07

Greatest Eclipse = 20:39:36.3 UT1
Penumbral Eclipse Magnitude = 0.8169
Umbral Eclipse Magnitude = -0.2610
Gamma = 1.1375
Axis = 1.0243°
Saros = 148 [6 of 70]
Ascending Node

Geocentric Coordinates at Greatest Eclipse

Sun	Moon
R.A. = 11h05m48.9s	R.A. = 23h03m51.6s
Dec. = +05°47'49.4"	Dec. = -04°53'44.4"
S.D. = 00°15'52.2"	S.D. = 00°14'43.4"
H.P. = 00°00'08.7"	H.P. = 00°54'02.2"

Eclipse Durations	Eclipse Contacts
Penumbral = 04h21m45s	P1 = 18:28:45 UT1
	P4 = 22:50:30 UT1

Parameters
Eph. = JPL DE430
Rule = Herald-Sinnott
ΔT = 95 s

www.EclipseWise.com   ©2020 F. Espenak

Partial Lunar Eclipse
2064 Feb 02

Greatest Eclipse = 21:47:22.4 UT1

Penumbral Eclipse Magnitude = 1.0264

Umbral Eclipse Magnitude = 0.0444

Gamma = 0.9969

Axis = 1.0078°

Saros = 115   [60 of 72]

Decending Node

Geocentric Coordinates at Greatest Eclipse

Sun	Moon
R.A. = 21h05m13.1s	R.A. = 09h06m49.3s
Dec. = -16°40'07.5"	Dec. = +17°36'03.4"
S.D. = 00°16'13.8"	S.D. = 00°16'31.7"
H.P. = 00°00'08.9"	H.P. = 01°00'39.5"

Eclipse Durations	Eclipse Contacts
Penumbral = 04h10m07s	P1 = 19:42:20 UT1
Umbral = 00h46m14s	U1 = 21:24:19 UT1
	U4 = 22:10:33 UT1
	P4 = 23:52:27 UT1

Parameters
Eph. = JPL DE430

Rule = Herald-Sinnott

ΔT = 95 s

www.EclipseWise.com   ©2020 F. Espenak

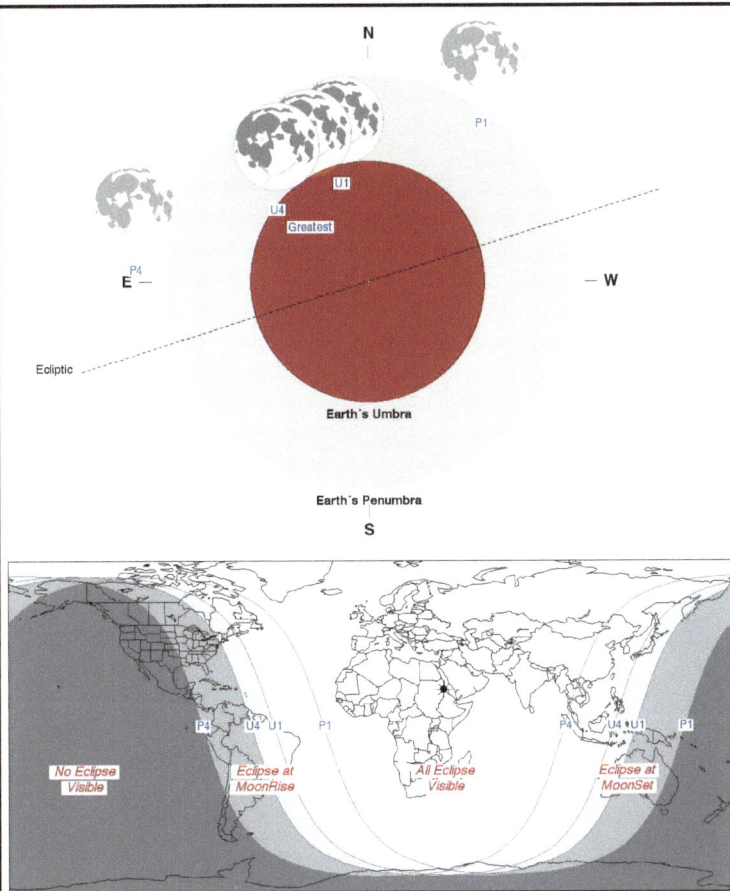

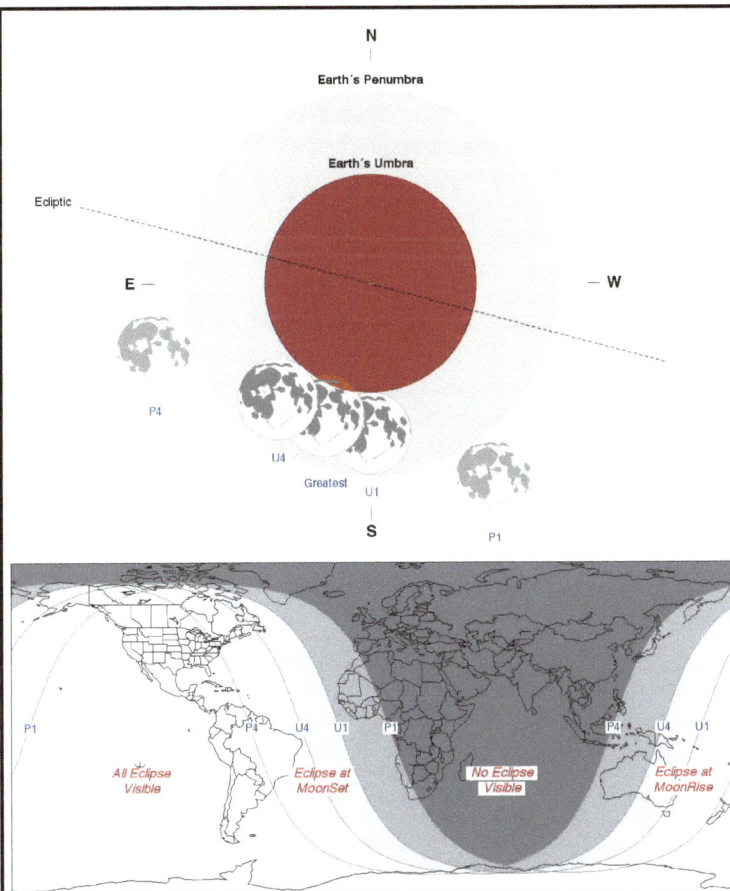

Partial Lunar Eclipse
2064 Jul 28

Greatest Eclipse = 07:51:12.9 UT1

Penumbral Eclipse Magnitude = 1.1428

Umbral Eclipse Magnitude = 0.1105

Gamma = -0.9473

Axis = 0.8840°

Saros = 120   [60 of 83]

Ascending Node

Geocentric Coordinates at Greatest Eclipse

Sun	Moon
R.A. = 08h33m43.4s	R.A. = 20h35m00.2s
Dec. = +18°45'12.2"	Dec. = -19°35'03.3"
S.D. = 00°15'45.1"	S.D. = 00°15'15.5"
H.P. = 00°00'08.7"	H.P. = 00°55'59.9"

Eclipse Durations	Eclipse Contacts
Penumbral = 04h45m24s	P1 = 05:28:29 UT1
Umbral = 01h18m11s	U1 = 07:12:02 UT1
	U4 = 08:30:13 UT1
	P4 = 10:13:53 UT1

Parameters
Eph. = JPL DE430

Rule = Herald-Sinnott

ΔT = 95 s

www.EclipseWise.com   ©2020 F. Espenak

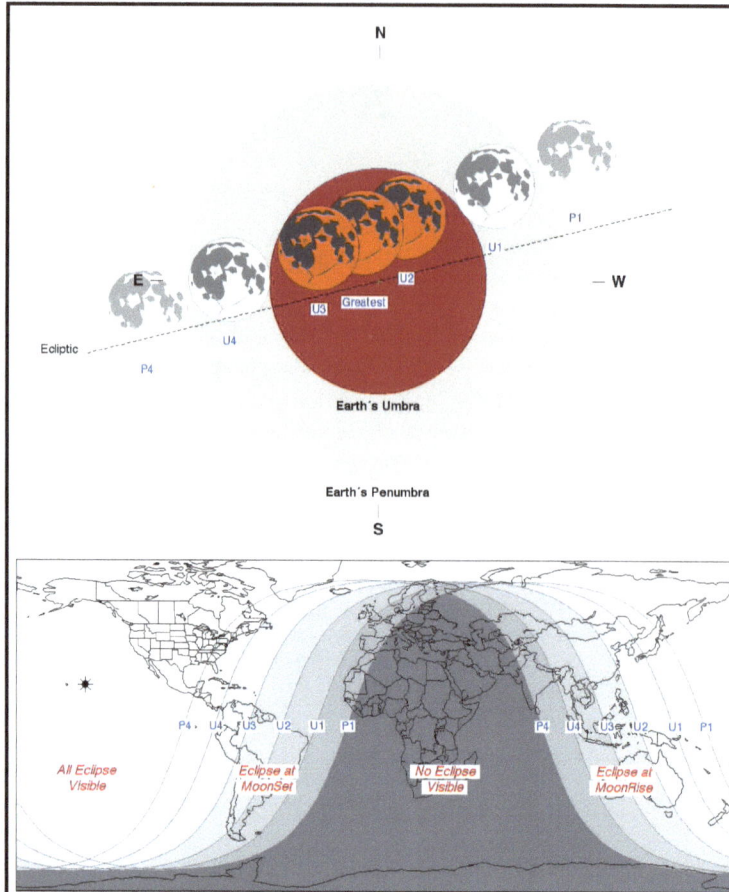

Total Lunar Eclipse
2065 Jan 22

Greatest Eclipse = 09:57:22.9 UT1
Penumbral Eclipse Magnitude = 2.2628
Umbral Eclipse Magnitude = 1.2298
Gamma = 0.3371
Axis = 0.3243°
Saros = 125 [51 of 72]
Decending Node

Geocentric Coordinates at Greatest Eclipse

Sun	Moon
R.A. = 20h20m52.4s	R.A. = 08h21m19.3s
Dec. = -19°29'52.6"	Dec. = +19°48'16.7"
S.D. = 00°16'15.1"	S.D. = 00°15'43.9"
H.P. = 00°00'08.9"	H.P. = 00°57'44.2"

Eclipse Durations	Eclipse Contacts
Penumbral = 05h39m06s	P1 = 07:07:47 UT1
Umbral = 03h29m52s	U1 = 08:12:29 UT1
Total = 01h09m57s	U2 = 09:22:26 UT1
	U3 = 10:32:23 UT1
	U4 = 11:42:20 UT1
Parameters	P4 = 12:46:53 UT1

Eph. = JPL DE430
Rule = Herald-Sinnott
ΔT = 96 s

www.EclipseWise.com   ©2020 F. Espenak

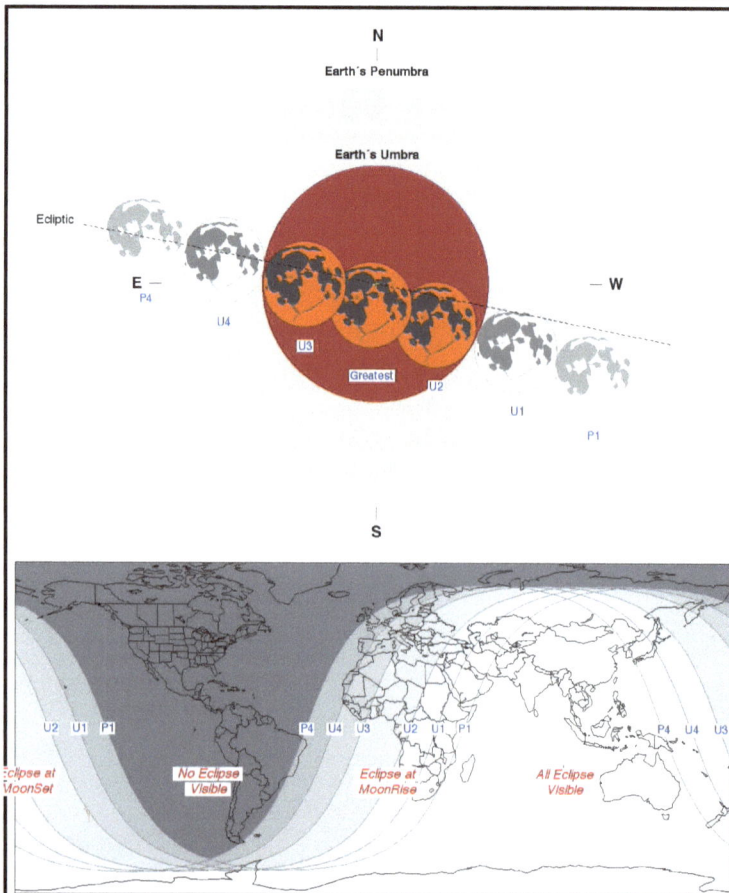

Total Lunar Eclipse
2065 Jul 17

Greatest Eclipse = 17:47:04.5 UT1
Penumbral Eclipse Magnitude = 2.5957
Umbral Eclipse Magnitude = 1.6188
Gamma = -0.1402
Axis = 0.1382°
Saros = 130 [37 of 71]
Ascending Node

Geocentric Coordinates at Greatest Eclipse

Sun	Moon
R.A. = 07h50m48.2s	R.A. = 19h50m58.2s
Dec. = +20°59'34.9"	Dec. = -21°07'32.3"
S.D. = 00°15'44.3"	S.D. = 00°16'06.6"
H.P. = 00°00'08.7"	H.P. = 00°59'07.5"

Eclipse Durations	Eclipse Contacts
Penumbral = 05h31m56s	P1 = 15:01:09 UT1
Umbral = 03h37m05s	U1 = 15:58:31 UT1
Total = 01h37m50s	U2 = 16:58:09 UT1
	U3 = 18:35:59 UT1
	U4 = 19:35:36 UT1
Parameters	P4 = 20:33:05 UT1

Eph. = JPL DE430
Rule = Herald-Sinnott
ΔT = 96 s

www.EclipseWise.com   ©2020 F. Espenak

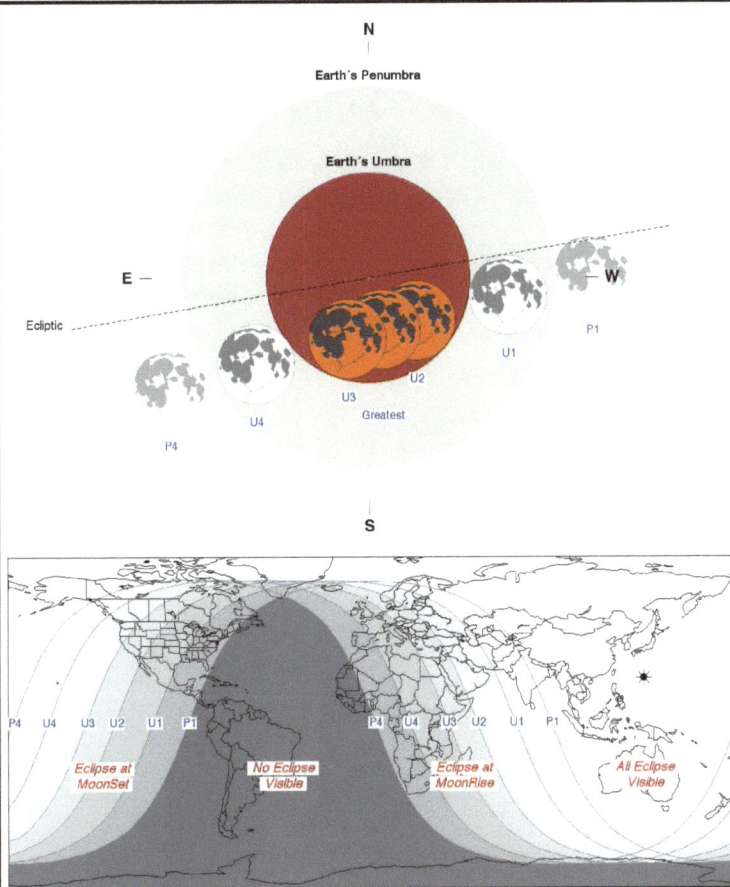

Total Lunar Eclipse
2066 Jan 11

Greatest Eclipse = 15:03:10.9 UT1
Penumbral Eclipse Magnitude = 2.2326
Umbral Eclipse Magnitude = 1.1445
Gamma = -0.3687
Axis = 0.3371°
Saros = 135 [26 of 71]
Decending Node

Geocentric Coordinates at Greatest Eclipse

Sun	Moon
R.A. = 19h33m47.1s	R.A. = 07h33m24.8s
Dec. = -21°40'59.0"	Dec. = +21°21'26.1"
S.D. = 00°16'15.8"	S.D. = 00°14'56.8"
H.P. = 00°00'08.9"	H.P. = 00°54'51.2"

Eclipse Durations	Eclipse Contacts
Penumbral = 06h01m42s	P1 = 12:02:18 UT1
Umbral = 03h36m07s	U1 = 13:15:09 UT1
Total = 00h59m27s	U2 = 14:33:28 UT1
	U3 = 15:32:56 UT1
	U4 = 16:51:16 UT1
	P4 = 18:04:00 UT1

Parameters
Eph. = JPL DE430
Rule = Herald-Sinnott
ΔT = 96 s

www.EclipseWise.com   ©2020 F. Espenak

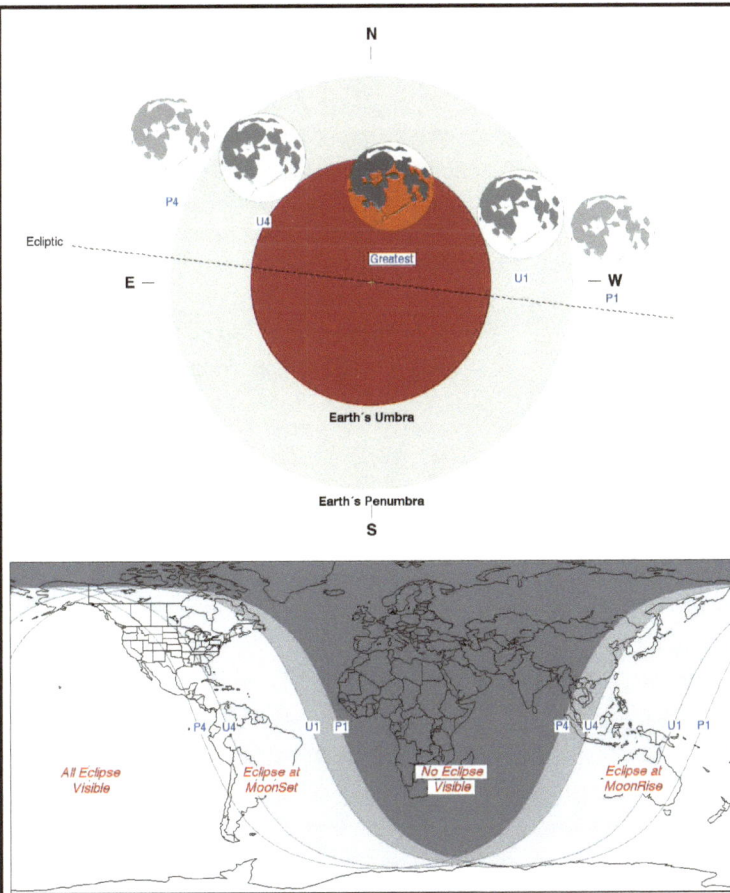

Partial Lunar Eclipse
2066 Jul 07

Greatest Eclipse = 09:28:52.6 UT1
Penumbral Eclipse Magnitude = 1.7246
Umbral Eclipse Magnitude = 0.7820
Gamma = 0.6056
Axis = 0.6181°
Saros = 140 [27 of 77]
Ascending Node

Geocentric Coordinates at Greatest Eclipse

Sun	Moon
R.A. = 07h07m48.8s	R.A. = 19h07m14.9s
Dec. = +22°30'58.3"	Dec. = -21°54'43.6"
S.D. = 00°15'43.9"	S.D. = 00°16'41.4"
H.P. = 00°00'08.7"	H.P. = 01°01'15.2"

Eclipse Durations	Eclipse Contacts
Penumbral = 04h53m16s	P1 = 07:02:15 UT1
Umbral = 02h52m17s	U1 = 08:02:44 UT1
	U4 = 10:55:00 UT1
	P4 = 11:55:31 UT1

Parameters
Eph. = JPL DE430
Rule = Herald-Sinnott
ΔT = 97 s

www.EclipseWise.com   ©2020 F. Espenak

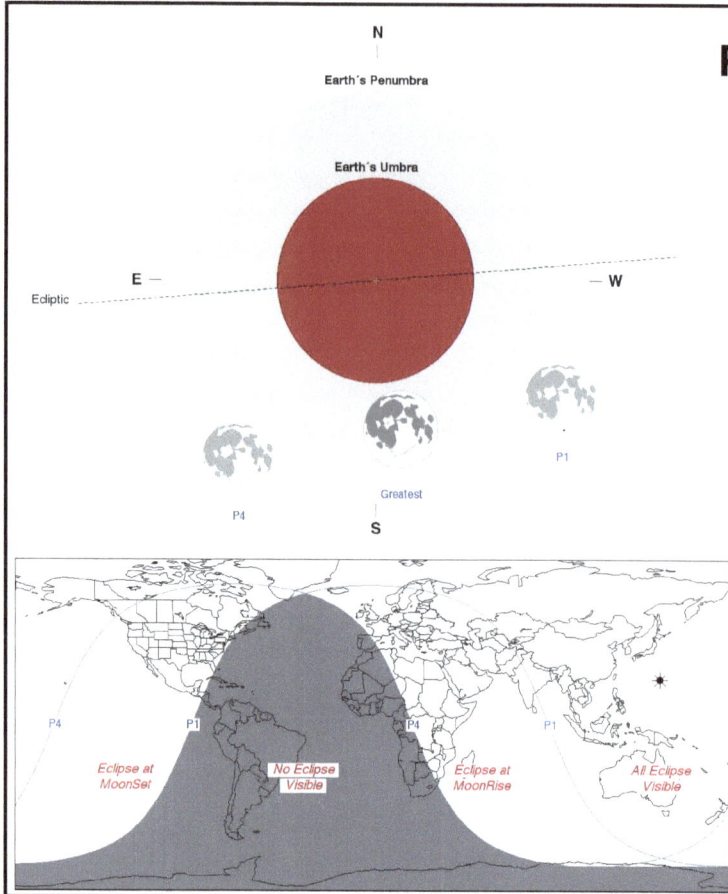

Penumbral Lunar Eclipse
2066 Dec 31

Greatest Eclipse = 14:28:32.4 UT1
Penumbral Eclipse Magnitude = 0.9840
Umbral Eclipse Magnitude = -0.1214
Gamma = -1.0540
Axis = 0.9485°
Saros = 145 [14 of 71]
Decending Node

Geocentric Coordinates at Greatest Eclipse

Sun	Moon
R.A. = 18h44m27.3s	R.A. = 06h43m44.2s
Dec. = -23°02'13.7"	Dec. = +22°06'11.2"
S.D. = 00°16'15.9"	S.D. = 00°14'42.9"
H.P. = 00°00'08.9"	H.P. = 00°54'00.3"

Eclipse Durations	Eclipse Contacts
Penumbral = 04h44m34s	P1 = 12:06:15 UT1
	P4 = 16:50:49 UT1

Parameters
Eph. = JPL DE430
Rule = Herald-Sinnott
ΔT = 97 s

www.EclipseWise.com   ©2020 F. Espenak

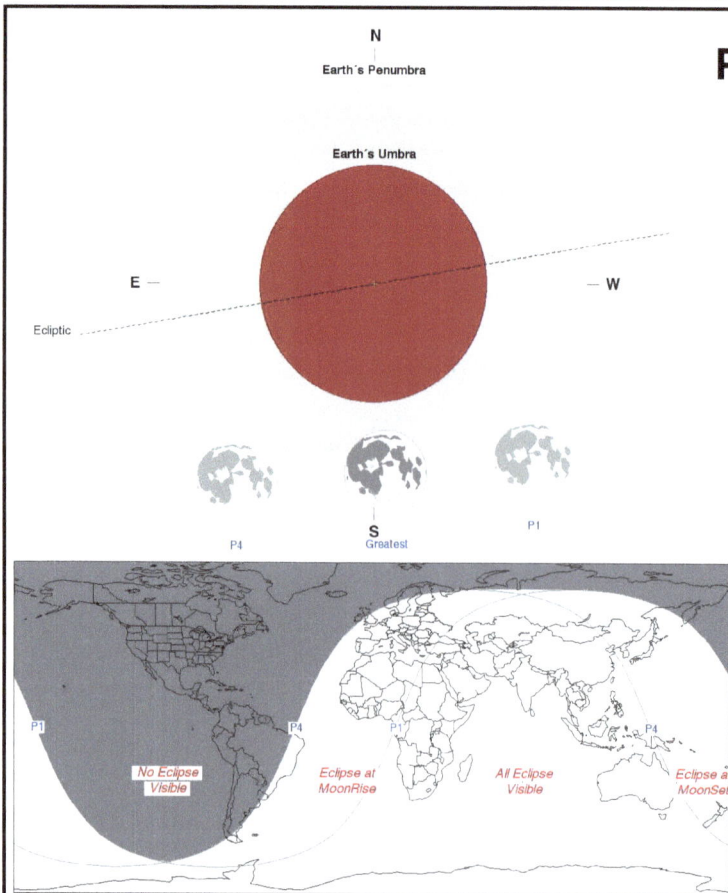

Penumbral Lunar Eclipse
2067 May 28

Greatest Eclipse = 18:54:29.9 UT1
Penumbral Eclipse Magnitude = 0.6469
Umbral Eclipse Magnitude = -0.3262
Gamma = -1.2013
Axis = 1.1917°
Saros = 112 [68 of 72]
Ascending Node

Geocentric Coordinates at Greatest Eclipse

Sun	Moon
R.A. = 04h23m02.8s	R.A. = 16h22m41.8s
Dec. = +21°33'42.4"	Dec. = -22°45'02.7"
S.D. = 00°15'47.0"	S.D. = 00°16'13.2"
H.P. = 00°00'08.7"	H.P. = 00°59'31.5"

Eclipse Durations	Eclipse Contacts
Penumbral = 03h29m49s	P1 = 17:09:39 UT1
	P4 = 20:39:29 UT1

Parameters
Eph. = JPL DE430
Rule = Herald-Sinnott
ΔT = 97 s

www.EclipseWise.com   ©2020 F. Espenak

Penumbral Lunar Eclipse
2067 Jun 27

Greatest Eclipse = 02:39:28.0 UT1

Penumbral Eclipse Magnitude = 0.3821

Umbral Eclipse Magnitude = -0.5685

Gamma = 1.3394

Axis = 1.3559°

Saros = 150 [4 of 71]

Ascending Node

Geocentric Coordinates at Greatest Eclipse

Sun	Moon
R.A. = 06h24m21.9s	R.A. = 18h23m33.9s
Dec. = +23°18'42.3"	Dec. = -21°58'06.1"
S.D. = 00°15'44.1"	S.D. = 00°16'33.1"
H.P. = 00°00'08.7"	H.P. = 01°00'44.6"

Eclipse Durations	Eclipse Contacts
Penumbral = 02h41m25s	P1 = 01:18:49 UT1
	P4 = 04:00:14 UT1

Parameters
Eph. = JPL DE430
Rule = Herald-Sinnott
ΔT = 97 s

Earth's Umbra

Earth's Penumbra

Eclipse at MoonRise

All Eclipse Visible

Eclipse at MoonSet

No Eclipse Visible

www.EclipseWise.com   ©2020 F. Espenak

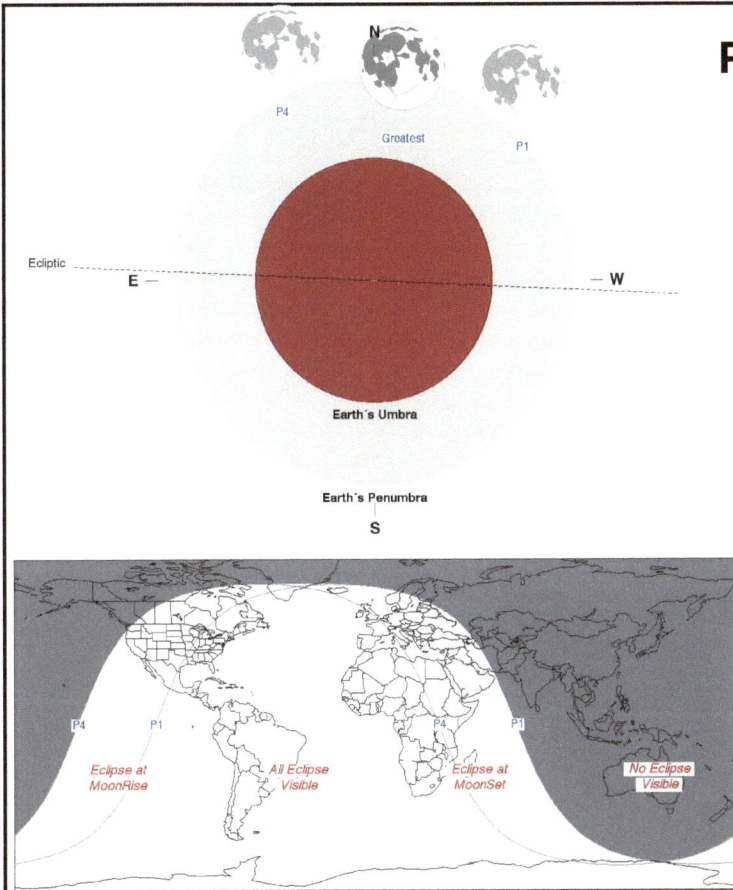

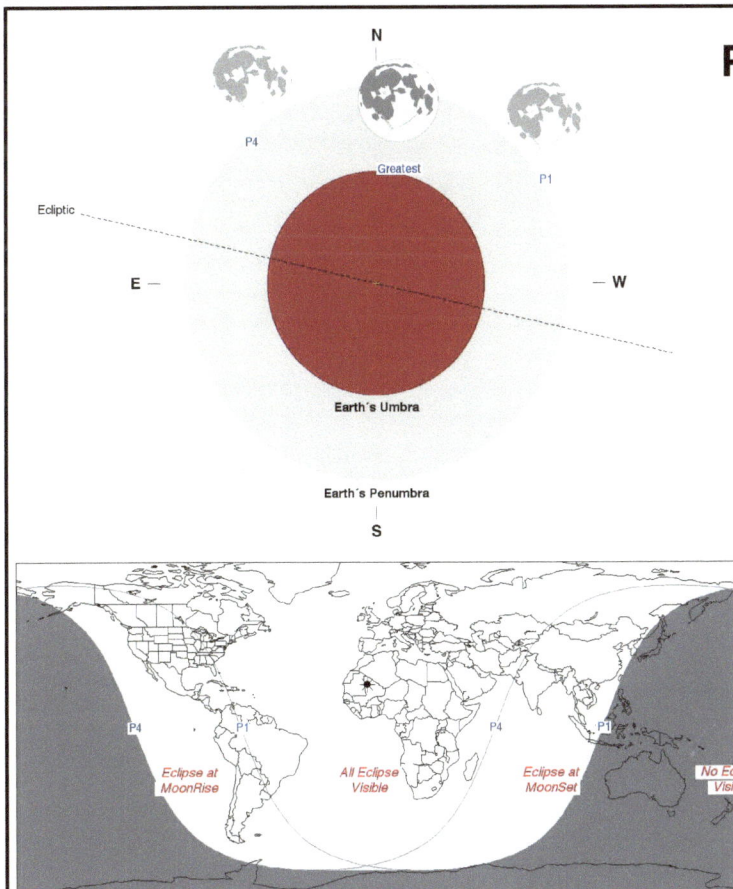

Penumbral Lunar Eclipse
2067 Nov 21

Greatest Eclipse = 00:03:03.8 UT1

Penumbral Eclipse Magnitude = 0.6611

Umbral Eclipse Magnitude = -0.3744

Gamma = 1.2107

Axis = 1.1575°

Saros = 117 [55 of 71]

Decending Node

Geocentric Coordinates at Greatest Eclipse

Sun	Moon
R.A. = 15h46m01.0s	R.A. = 03h45m24.1s
Dec. = -19°52'08.2"	Dec. = +21°01'03.2"
S.D. = 00°16'11.2"	S.D. = 00°15'37.9"
H.P. = 00°00'08.9"	H.P. = 00°57'22.2"

Eclipse Durations	Eclipse Contacts
Penumbral = 03h42m52s	P1 = 22:11:33 UT1
	P4 = 01:54:25 UT1

Parameters
Eph. = JPL DE430
Rule = Herald-Sinnott
ΔT = 98 s

Earth's Umbra

Earth's Penumbra

Eclipse at MoonRise

All Eclipse Visible

Eclipse at MoonSet

No Eclipse Visible

www.EclipseWise.com   ©2020 F. Espenak

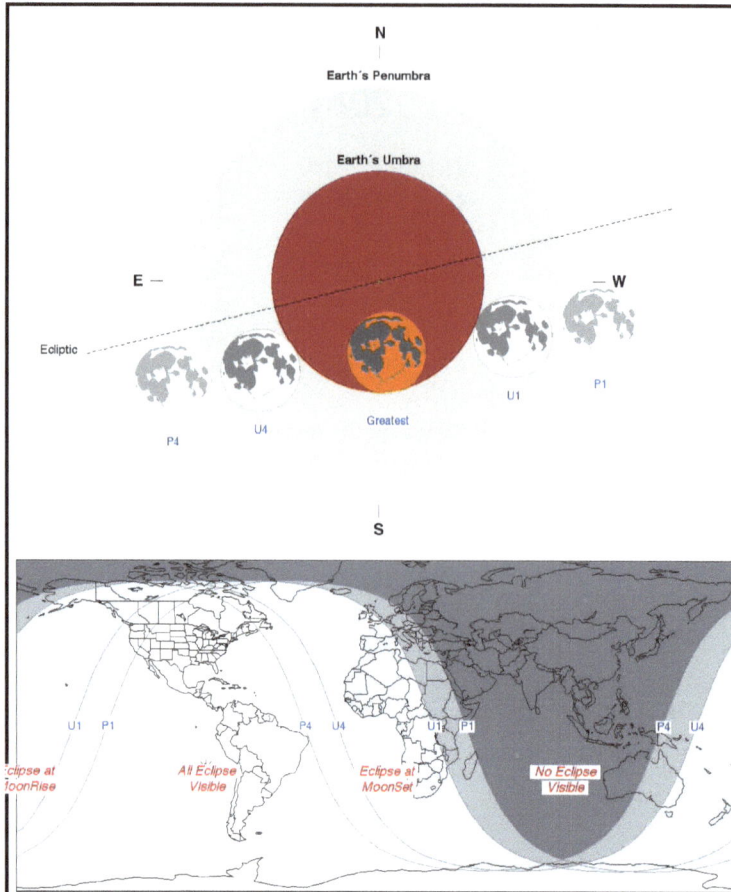

Partial Lunar Eclipse
2068 May 17

Greatest Eclipse = 05:40:38.8 UT1
Penumbral Eclipse Magnitude = 1.9893
Umbral Eclipse Magnitude = 0.9599
Gamma = -0.4852
Axis = 0.4559°
Saros = 122 [59 of 74]
Ascending Node

Geocentric Coordinates at Greatest Eclipse

Sun	Moon
R.A. = 03h39m33.2s	R.A. = 15h39m17.9s
Dec. = +19°31'07.8"	Dec. = -19°58'14.9"
S.D. = 00°15'49.0"	S.D. = 00°15'21.9"
H.P. = 00°00'08.7"	H.P. = 00°56'23.4"

Eclipse Durations	Eclipse Contacts
Penumbral = 05h37m37s	P1 = 02:51:48 UT1
Umbral = 03h19m57s	U1 = 04:00:43 UT1
	U4 = 07:20:40 UT1
	P4 = 08:29:25 UT1

Parameters
Eph. = JPL DE430
Rule = Herald-Sinnott
ΔT = 98 s

www.EclipseWise.com   ©2020 F. Espenak

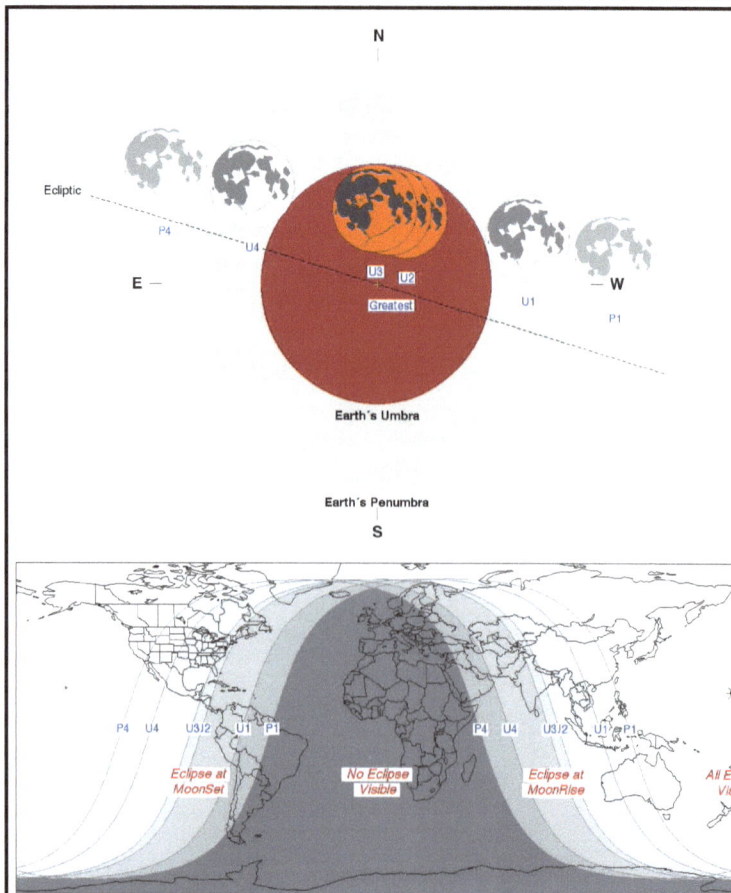

Total Lunar Eclipse
2068 Nov 09

Greatest Eclipse = 11:45:21.2 UT1
Penumbral Eclipse Magnitude = 2.0028
Umbral Eclipse Magnitude = 1.0216
Gamma = 0.4645
Axis = 0.4675°
Saros = 127 [45 of 72]
Decending Node

Geocentric Coordinates at Greatest Eclipse

Sun	Moon
R.A. = 15h01m47.2s	R.A. = 03h01m25.7s
Dec. = -17°09'37.3"	Dec. = +17°37'12.0"
S.D. = 00°16'08.8"	S.D. = 00°16'27.4"
H.P. = 00°00'08.9"	H.P. = 01°00'23.7"

Eclipse Durations	Eclipse Contacts
Penumbral = 05h12m04s	P1 = 09:09:20 UT1
Umbral = 03h11m07s	U1 = 10:09:46 UT1
Total = 00h22m07s	U2 = 11:34:16 UT1
	U3 = 11:56:24 UT1
	U4 = 13:20:53 UT1
	P4 = 14:21:24 UT1

Parameters
Eph. = JPL DE430
Rule = Herald-Sinnott
ΔT = 99 s

www.EclipseWise.com   ©2020 F. Espenak

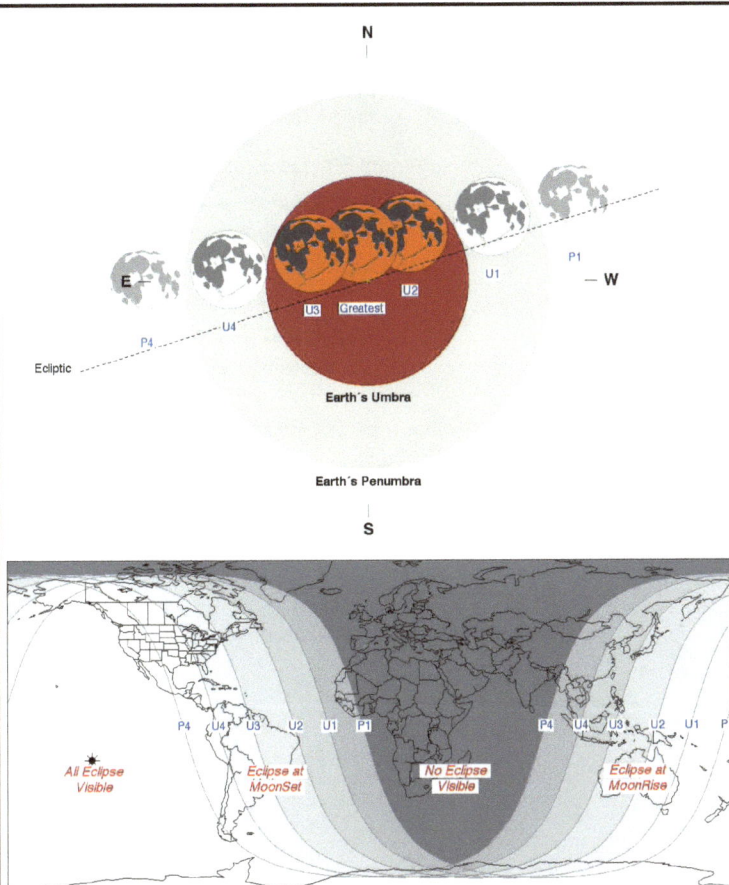

Total Lunar Eclipse
2069 May 06

Greatest Eclipse = 09:08:17.6 UT1
Penumbral Eclipse Magnitude = 2.4032
Umbral Eclipse Magnitude = 1.3296
Gamma = 0.2717
Axis = 0.2454°
Saros = 132 [33 of 71]
Ascending Node

Geocentric Coordinates at Greatest Eclipse

Sun	Moon
R.A. = 02h55m56.2s	R.A. = 14h56m07.8s
Dec. = +16°44'53.2"	Dec. = -16°30'25.4"
S.D. = 00°15'51.4"	S.D. = 00°14'46.2"
H.P. = 00°00'08.7"	H.P. = 00°54'12.3"

Eclipse Durations	Eclipse Contacts
Penumbral = 06h09m07s	P1 = 06:03:43 UT1
Umbral = 03h47m06s	U1 = 07:14:45 UT1
Total = 01h25m23s	U2 = 08:25:37 UT1
	U3 = 09:51:00 UT1
	U4 = 11:01:51 UT1
	P4 = 12:12:50 UT1

Parameters
Eph. = JPL DE430
Rule = Herald-Sinnott
ΔT = 99 s

www.EclipseWise.com   ©2020 F. Espenak

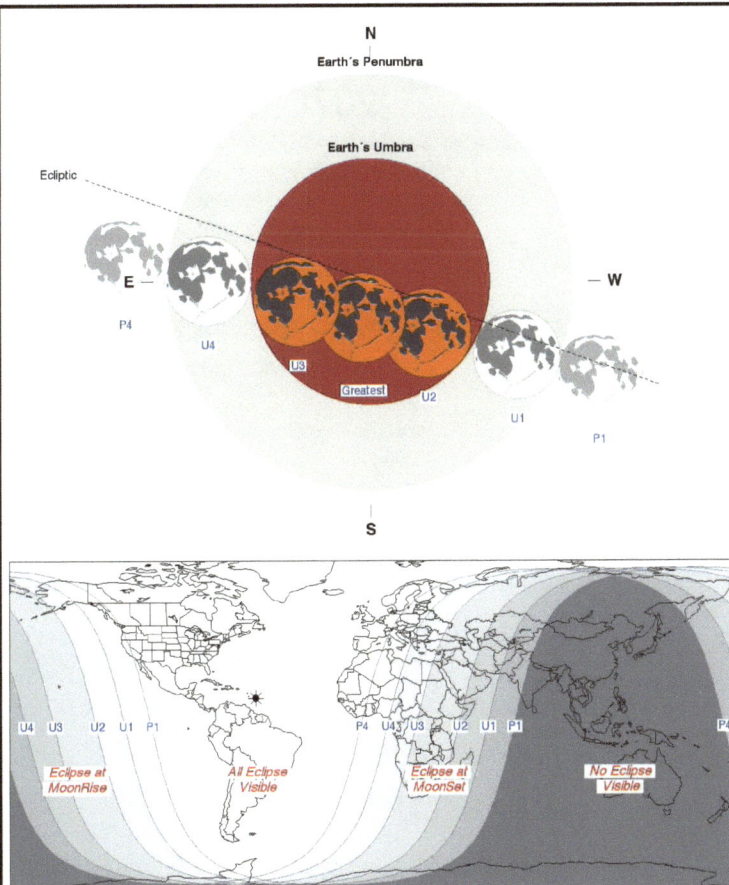

Total Lunar Eclipse
2069 Oct 30

Greatest Eclipse = 03:33:26.4 UT1
Penumbral Eclipse Magnitude = 2.4302
Umbral Eclipse Magnitude = 1.4683
Gamma = -0.2263
Axis = 0.2317°
Saros = 137 [29 of 78]
Decending Node

Geocentric Coordinates at Greatest Eclipse

Sun	Moon
R.A. = 14h19m49.6s	R.A. = 02h20m02.8s
Dec. = -13°56'35.4"	Dec. = +13°43'03.8"
S.D. = 00°16'06.2"	S.D. = 00°16'44.5"
H.P. = 00°00'08.9"	H.P. = 01°01'26.5"

Eclipse Durations	Eclipse Contacts
Penumbral = 05h16m18s	P1 = 00:55:17 UT1
Umbral = 03h26m25s	U1 = 01:50:14 UT1
Total = 01h27m37s	U2 = 02:49:38 UT1
	U3 = 04:17:15 UT1
	U4 = 05:16:39 UT1
	P4 = 06:11:35 UT1

Parameters
Eph. = JPL DE430
Rule = Herald-Sinnott
ΔT = 99 s

www.EclipseWise.com   ©2020 F. Espenak

Penumbral Lunar Eclipse
2070 Apr 25

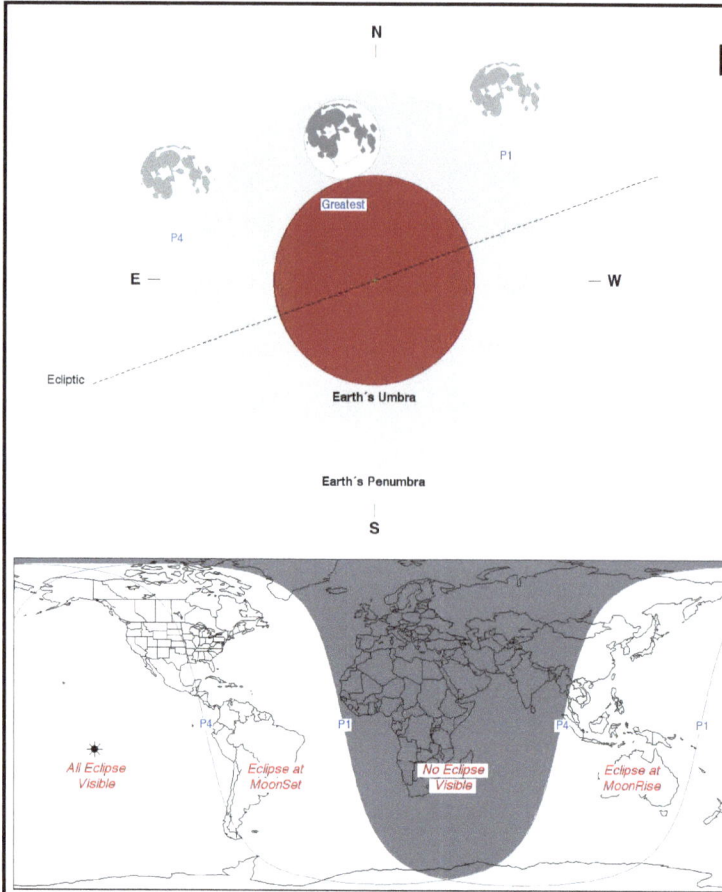

Greatest Eclipse = 09:19:44.8 UT1
Penumbral Eclipse Magnitude = 1.0582
Umbral Eclipse Magnitude = -0.0142
Gamma = 1.0044
Axis = 0.9109°
Saros = 142 [21 of 73]
Ascending Node

Geocentric Coordinates at Greatest Eclipse

Sun	Moon
R.A. = 02h12m57.7s	R.A. = 14h13m51.0s
Dec. = +13°21'41.4"	Dec. = -12°28'35.7"
S.D. = 00°15'54.1"	S.D. = 00°14'49.7"
H.P. = 00°00'08.7"	H.P. = 00°54'25.3"

Eclipse Durations	Eclipse Contacts
Penumbral = 04h48m08s	P1 = 06:55:40 UT1
	P4 = 11:43:47 UT1

Parameters
Eph. = JPL DE430
Rule = Herald-Sinnott
ΔT = 100 s

www.EclipseWise.com   ©2020 F. Espenak

Partial Lunar Eclipse
2070 Oct 19

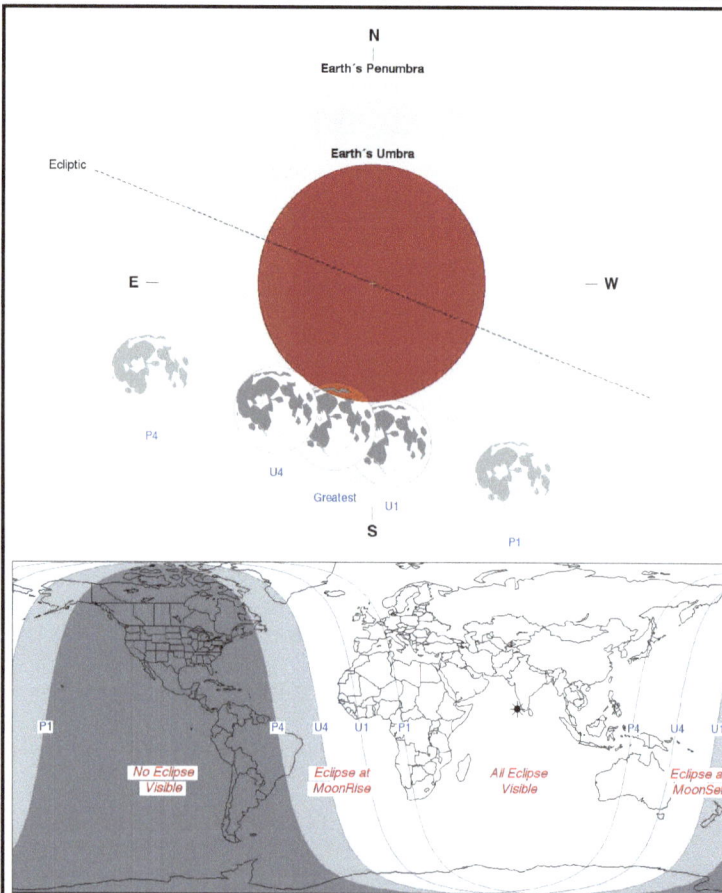

Greatest Eclipse = 18:49:31.8 UT1
Penumbral Eclipse Magnitude = 1.1325
Umbral Eclipse Magnitude = 0.1450
Gamma = -0.9406
Axis = 0.9354°
Saros = 147 [11 of 70]
Decending Node

Geocentric Coordinates at Greatest Eclipse

Sun	Moon
R.A. = 13h39m13.7s	R.A. = 01h40m14.8s
Dec. = -10°18'14.5"	Dec. = +09°24'10.4"
S.D. = 00°16'03.4"	S.D. = 00°16'15.6"
H.P. = 00°00'08.8"	H.P. = 00°59'40.4"

Eclipse Durations	Eclipse Contacts
Penumbral = 04h24m14s	P1 = 16:37:27 UT1
Umbral = 01h23m44s	U1 = 18:07:44 UT1
	U4 = 19:31:29 UT1
	P4 = 21:01:40 UT1

Parameters
Eph. = JPL DE430
Rule = Herald-Sinnott
ΔT = 100 s

www.EclipseWise.com   ©2020 F. Espenak

Key to Catalog of Lunar Eclipses

The following catalog lists all lunar eclipses during a 25-year period.

A brief description of each parameter in the catalog appears below.

Date — Gregorian date of Greatest Eclipse

Greatest Eclipse — Universal Time of Greatest Eclipse

Saros — Saros Series Number of the eclipse

Type — Lunar Eclipse Type

> N = Penumbral Lunar Eclipse
> P = Partial Lunar Eclipse
> T = Total Lunar Eclipse

> + = Central total eclipse (Moon's center passes north of shadow axis)
> – = Central total eclipse (Moon's center passes south of shadow axis)

> b = Saros series begins (first penumbral eclipse in a Saros series)
> e = Saros series ends (last penumbral eclipse in a Saros series)

Gamma — minimum distance from center of the Moon to axis of Earth's umbral shadow

P. Mag — Penumbral Magnitude; fraction of the Moon's diameter immersed in the penumbra

U. Mag — Umbral Magnitude; fraction of the Moon's diameter immersed in the umbra

Phase Durations
> **Pen** — elapsed time from contact P1 to P4 (minutes)
> **Partial** — elapsed time from contact U1 to U4 (minutes)
> **Total** — elapsed time from contact U2 to U3 (minutes)

Lat & Long — latitude and longitude where the Moon appears in the zenith at Greatest Eclipse

Photo 2–8 Time sequence of the total lunar eclipse of 2014 April 15. ©2014 F. Espenak

Catalog of Lunar Eclipses: 2061 to 2085

Date	Greatest Eclipse	Saros	Type	Gamma	P. Mag	U.Mag	Pen	Partial	Total	Lat.	Long.
2061 Apr 04	21:52:32	123	T	0.4300	2.1064	1.0360	355.8	210.3	30.9	5.8S	32.7E
2061 Sep 29	9:36:40	128	T	-0.3810	2.1576	1.1640	325.5	203.1	59.6	2.4N	146.4W
2062 Mar 25	3:32:17	133	T	-0.3150	2.2925	1.2715	339.1	212.0	75.3	2.2S	51.7W
2062 Sep 18	18:32:28	138	T	0.3736	2.1979	1.1515	350.0	213.1	60.2	1.2S	80.2E
2063 Mar 14	16:04:14	143	P	-1.0008	1.0108	0.0363	248.5	41.9	-	1.5N	120.7E
2063 Sep 07	20:39:36	148	N	1.1375	0.8121	-0.2657	261.2	-	-	4.9S	49.1E
2064 Feb 02	21:47:22	115	P	0.9969	1.0215	0.0395	249.8	43.7	-	17.6N	37.0E
2064 Jul 28	7:51:13	120	P	-0.9473	1.1378	0.1055	285.0	76.5	-	19.6S	115.8W
2065 Jan 22	9:57:23	125	T	0.3371	2.2579	1.2248	339.0	209.7	69.4	19.8N	146.3W
2065 Jul 17	17:47:04	130	T-	-0.1402	2.5907	1.6138	331.9	217.0	97.7	21.1S	94.9E
2066 Jan 11	15:03:11	135	T	-0.3687	2.2276	1.1395	361.6	216.0	58.7	21.4N	136.1E
2066 Jul 07	9:28:53	140	P	0.6056	1.7196	0.7770	293.1	172.0	-	21.9S	141.1W
2066 Dec 31	14:28:32	145	N	-1.0540	0.9789	-0.1264	284.2	-	-	22.1N	143.5E
2067 May 28	18:54:30	112	N	-1.2013	0.6416	-0.3316	209.2	-	-	22.8S	75.7E
2067 Jun 27	2:39:28	150	N	1.3394	0.3770	-0.5736	160.5	-	-	22.0S	39.3W
2067 Nov 21	0:03:04	117	N	1.2107	0.6557	-0.3798	222.2	-	-	21.0N	4.5W
2068 May 17	5:40:39	122	P	-0.4852	1.9839	0.9545	337.5	199.8	-	20.0S	86.1W
2068 Nov 09	11:45:21	127	T	0.4645	1.9974	1.0161	312.0	190.9	19.2	17.6N	179.5E
2069 May 06	9:08:18	132	T+	0.2717	2.3977	1.3242	369.1	227.0	85.0	16.5S	137.9W
2069 Oct 30	3:33:26	137	T-	-0.2263	2.4247	1.4628	316.3	206.4	87.4	13.7N	57.4W
2070 Apr 25	9:19:45	142	Nx	1.0044	1.0527	-0.0197	287.7	-	-	12.5S	140.2W
2070 Oct 19	18:49:32	147	P	-0.9406	1.1270	0.1395	263.9	82.3	-	9.4N	74.1E
2071 Mar 16	1:29:28	114	N	-1.0757	0.8890	-0.1183	245.8	-	-	0.8N	20.5W
2071 Sep 09	15:03:59	119	N	1.0835	0.9000	-0.1575	265.9	-	-	4.2S	133.0E
2072 Mar 04	15:21:26	124	T	-0.3431	2.2137	1.2452	313.9	200.1	69.0	5.6N	132.4E
2072 Aug 28	16:04:01	129	T	0.3563	2.2439	1.1673	366.9	221.1	64.8	8.9S	119.2E
2073 Feb 22	7:23:11	134	T	0.3389	2.2230	1.2514	314.6	200.4	69.8	10.3N	107.4W
2073 Aug 17	17:40:59	139	T	-0.3998	2.1490	1.1024	351.3	212.3	50.8	13.4S	95.8E
2074 Feb 11	20:54:15	144	N	1.0612	0.9203	-0.0960	250.2	-	-	14.7N	50.2E
2074 Jul 08	17:19:54	111	N	1.4457	0.1884	-0.7751	117.3	-	-	20.9S	101.3E
2074 Aug 07	1:54:21	149	N	-1.1291	0.7826	-0.2079	232.9	-	-	17.4S	26.9W
2075 Jan 02	9:53:20	116	N	-1.1643	0.7729	-0.3256	255.8	-	-	21.8N	147.3W
2075 Jun 28	9:53:52	121	P	0.6897	1.5639	0.6235	284.2	157.6	-	22.5S	147.6W
2075 Dec 22	8:54:11	126	P	-0.4945	2.0024	0.9028	358.6	203.4	-	23.0N	133.9W
2076 Jun 17	2:38:02	131	Tm	-0.0452	2.7570	1.7959	326.2	215.9	100.9	23.4S	39.2W
2076 Dec 10	11:33:06	136	T+	0.2102	2.5006	1.4476	353.1	221.4	91.6	23.2N	175.0W
2077 Jun 06	14:58:06	141	P	-0.8388	1.3274	0.3139	294.4	125.7	-	23.5S	135.0E
2077 Nov 29	21:34:07	146	P	0.8855	1.2326	0.2372	272.8	105.6	-	22.5N	33.4E
2078 Apr 27	4:33:59	113	N	1.2223	0.6577	-0.4227	239.0	-	-	13.0S	68.6W
2078 Oct 21	3:06:17	118	N	-1.1022	0.8191	-0.1442	225.5	-	-	9.8N	49.9W
2078 Nov 19	12:38:15	156	N	1.5148	0.0633	-0.9028	67.2	-	-	21.1N	166.3E
2079 Apr 16	5:08:58	123	P	0.4800	2.0119	0.9471	350.9	204.1	-	9.8S	77.1W
2079 Oct 10	17:28:42	128	T	-0.4246	2.0806	1.0811	324.5	199.3	43.1	6.5N	94.7E
2080 Apr 04	11:21:50	133	T	-0.2751	2.3626	1.3479	339.0	214.2	82.8	6.4S	169.9W
2080 Sep 29	1:50:53	138	T	0.3203	2.2986	1.2462	354.5	218.0	74.4	3.0N	30.3W
2081 Mar 25	0:20:12	143	P	-0.9688	1.0673	0.0973	253.1	68.0	-	2.9S	4.1W
2081 Sep 18	3:33:36	148	N	1.0747	0.9291	-0.1524	276.4	-	-	0.8S	55.3W
2082 Feb 13	6:27:30	115	P	1.0101	0.9974	0.0153	247.9	27.3	-	14.1N	92.9W
2082 Aug 08	14:44:52	120	Nx	-1.0204	1.0030	-0.0275	270.5	-	-	16.8S	140.6E
2083 Feb 02	18:24:55	125	T	0.3464	2.2418	1.2070	339.7	209.5	67.2	16.9N	87.3E
2083 Jul 29	1:03:43	130	T-	-0.2143	2.4537	1.4791	328.8	213.6	91.1	18.9S	14.2W
2084 Jan 22	23:11:09	135	T	-0.3610	2.2425	1.1531	362.9	217.0	61.3	19.2N	15.0E
2084 Jul 17	16:56:59	140	P	0.5313	1.8557	0.9136	298.9	182.1	-	20.4S	107.2E
2085 Jan 10	22:30:37	145	N	-1.0453	0.9944	-0.1102	285.7	-	-	20.8N	24.1E
2085 Jun 08	2:15:43	112	N	-1.2746	0.5079	-0.4668	189.2	-	-	24.2S	34.1W
2085 Jul 07	10:02:46	150	N	1.2695	0.5064	-0.4461	184.2	-	-	21.2S	149.7W
2085 Dec 01	8:23:42	117	N	1.2190	0.6400	-0.3944	219.2	-	-	23.1N	128.7W

Section 3: Phases of the Moon: 2061 to 2070

Photo 3–1 Various phases of the Moon over one lunar month. ©2010 F. Espenak

Phases of the Moon

Year	New Moon	First Quarter	Full Moon	Last Quarter
2061			Jan 06 02:24	Jan 13 13:57
	Jan 21 15:16	Jan 28 18:10	Feb 04 15:22	Feb 12 11:52
	Feb 20 05:31	Feb 27 01:51	Mar 06 05:54	Mar 14 08:31
	Mar 21 17:23	Mar 28 09:26	Apr 04 21:47 t	Apr 13 02:10
	Apr 20 03:04 T	Apr 26 17:55	May 04 14:13	May 12 16:10
	May 19 11:03	May 26 04:12	Jun 03 06:09	Jun 11 02:42
	Jun 17 18:03	Jun 24 16:54	Jul 02 20:52	Jul 10 10:23
	Jul 17 01:10	Jul 24 08:05	Aug 01 10:11	Aug 08 16:09
	Aug 15 09:39	Aug 23 01:18	Aug 30 22:18	Sep 06 21:12
	Sep 13 20:37	Sep 21 19:44	Sep 29 09:32 t	Oct 06 02:57
	Oct 13 10:41 A	Oct 21 14:24	Oct 28 20:12	Nov 04 10:53
	Nov 12 03:40	Nov 20 08:11	Nov 27 06:32	Dec 03 22:12
	Dec 11 22:32	Dec 19 23:58	Dec 26 16:53	

Year	New Moon	First Quarter	Full Moon	Last Quarter
2062				Jan 02 13:21
	Jan 10 17:52	Jan 18 12:51	Jan 25 03:37	Feb 01 07:43
	Feb 09 12:11	Feb 16 22:38	Feb 23 15:08	Mar 03 03:49
	Mar 11 04:13 P	Mar 18 05:58	Mar 25 03:35 t	Apr 01 23:55
	Apr 09 17:17	Apr 16 12:03	Apr 23 16:57	May 01 18:33
	May 09 03:22	May 15 18:17	May 23 07:03	May 31 10:44
	Jun 07 11:12	Jun 14 01:53	Jun 21 21:43	Jun 29 23:54
	Jul 06 17:53	Jul 13 11:43	Jul 21 12:47	Jul 29 10:04
	Aug 05 00:40	Aug 12 00:21	Aug 20 03:55	Aug 27 17:49
	Sep 03 08:42 P	Sep 10 15:59	Sep 18 18:36 t	Sep 26 00:11
	Oct 02 18:49	Oct 10 10:27	Oct 18 08:18	Oct 25 06:28
	Nov 01 07:32	Nov 09 06:50	Nov 16 20:48	Nov 23 13:58
	Nov 30 23:01	Dec 09 03:28	Dec 16 08:17	Dec 22 23:40
	Dec 30 16:57			

All times are in Universal Time (UT1).
Eclipses at New Moon (solar eclipse) or Full Moon (lunar eclipse), are indicated by these symbols:

Solar Eclipses	Lunar Eclipses
T – Total	t – Total (Umbral)
A – Annular	p – Partial (Umbral)
H – Hybrid	n – Penumbral
P – Partial	

Phases of the Moon

Year	New Moon	First Quarter	Full Moon	Last Quarter
2063		Jan 07 22:16	Jan 14 19:11	Jan 21 12:05
	Jan 29 12:23	Feb 06 13:37	Feb 13 05:48	Feb 20 03:07
	Feb 28 07:38 A	Mar 08 01:06	Mar 14 16:14 p	Mar 21 20:16
	Mar 30 00:50	Apr 06 09:18	Apr 13 02:34	Apr 20 14:42
	Apr 28 14:52	May 05 15:20	May 12 13:11	May 20 09:16
	May 28 01:47	Jun 03 20:28	Jun 11 00:43	Jun 19 02:43
	Jun 26 10:25	Jul 03 02:01	Jul 10 13:48	Jul 18 18:05
	Jul 25 17:55	Aug 01 09:09	Aug 09 04:40	Aug 17 07:01
	Aug 24 01:17 T	Aug 30 19:04	Sep 07 20:53 n	Sep 15 17:44
	Sep 22 09:21	Sep 29 08:39	Oct 07 13:27	Oct 15 02:49
	Oct 21 18:46	Oct 29 02:13	Nov 06 05:22	Nov 13 10:56
	Nov 20 06:09	Nov 27 22:59	Dec 05 20:06	Dec 12 18:49
	Dec 19 20:04	Dec 27 20:57		

Year	New Moon	First Quarter	Full Moon	Last Quarter
2064			Jan 04 09:31	Jan 11 03:14
	Jan 18 12:37	Jan 26 17:42	Feb 02 21:37 p	Feb 09 12:55
	Feb 17 07:03 A	Feb 25 11:23	Mar 03 08:19	Mar 10 00:33
	Mar 18 01:45	Mar 26 01:13	Apr 01 17:40	Apr 08 14:25
	Apr 16 19:02	Apr 24 11:17	May 01 02:08	May 08 06:16
	May 16 09:55	May 23 18:15	May 30 10:36	Jun 06 23:23
	Jun 14 22:20	Jun 21 23:13	Jun 28 20:08	Jul 06 16:48
	Jul 14 08:46	Jul 21 03:35	Jul 28 07:40 p	Aug 05 09:41
	Aug 12 17:49 T	Aug 19 08:55	Aug 26 21:35	Sep 04 01:29
	Sep 11 02:11	Sep 17 16:45	Sep 25 13:38	Oct 03 15:50
	Oct 10 10:34	Oct 17 04:22	Oct 25 07:06	Nov 02 04:24
	Nov 08 19:45	Nov 15 20:14	Nov 24 00:58	Dec 01 15:00
	Dec 08 06:28	Dec 15 15:45	Dec 23 18:14	Dec 30 23:50

Year	New Moon	First Quarter	Full Moon	Last Quarter
2065	Jan 06 19:15	Jan 14 13:19	Jan 22 09:53 t	Jan 29 07:38
	Feb 05 10:02 P	Feb 13 10:50	Feb 20 23:11	Feb 27 15:29
	Mar 07 02:15	Mar 15 06:25	Mar 22 09:56	Mar 29 00:24
	Apr 05 19:01	Apr 13 22:38	Apr 20 18:36	Apr 27 11:02
	May 05 11:30	May 13 10:52	May 20 02:05	May 26 23:38
	Jun 04 03:05	Jun 11 19:25	Jun 18 09:28	Jun 25 14:08
	Jul 03 17:16 P	Jul 11 01:16	Jul 17 17:45 t	Jul 25 06:22
	Aug 02 05:46 P	Aug 09 05:52	Aug 16 03:45	Aug 23 23:56
	Aug 31 16:39	Sep 07 10:49	Sep 14 16:05	Sep 22 18:09
	Sep 30 02:24	Oct 06 17:37	Oct 14 07:04	Oct 22 11:53
	Oct 29 11:48	Nov 05 03:26	Nov 13 00:37	Nov 21 03:51
	Nov 27 21:40	Dec 04 16:54	Dec 12 19:52	Dec 20 17:12
	Dec 27 08:27 P			

All times are in Universal Time (UT1).
Eclipses at New Moon (solar eclipse) or Full Moon (lunar eclipse), are indicated by these symbols:

Solar Eclipses	Lunar Eclipses
T - Total	t - Total (Umbral)
A - Annular	p - Partial (Umbral)
H - Hybrid	n - Penumbral
P - Partial	

Phases of the Moon

Year	New Moon	First Quarter	Full Moon	Last Quarter
2066		Jan 03 09:56	Jan 11 15:07 t	Jan 19 03:48
	Jan 25 20:14	Feb 02 05:44	Feb 10 08:29	Feb 17 12:14
	Feb 24 08:50	Mar 04 02:48	Mar 11 22:48	Mar 18 19:25
	Mar 25 22:13	Apr 02 23:09	Apr 10 10:03	Apr 17 02:23
	Apr 24 12:29	May 02 16:57	May 09 18:58	May 16 10:01
	May 24 03:38	Jun 01 07:13	Jun 08 02:31	Jun 14 19:10
	Jun 22 19:15 A	Jun 30 17:59	Jul 07 09:34 p	Jul 14 06:38
	Jul 22 10:34	Jul 30 02:01	Aug 05 16:59	Aug 12 20:59
	Aug 21 00:50	Aug 28 08:25	Sep 04 01:37	Sep 11 14:16
	Sep 19 13:47	Sep 26 14:19	Oct 03 12:25	Oct 11 09:43
	Oct 19 01:42	Oct 25 20:52	Nov 02 02:13	Nov 10 05:45
	Nov 17 13:06	Nov 24 05:10	Dec 01 19:16	Dec 10 00:38
	Dec 17 00:17 T	Dec 23 16:07	Dec 31 14:41 n	

Year	New Moon	First Quarter	Full Moon	Last Quarter
2067				Jan 08 17:01
	Jan 15 11:17	Jan 22 06:17	Jan 30 10:29	Feb 07 06:14
	Feb 13 21:57	Feb 20 23:30	Mar 01 04:42	Mar 08 16:16
	Mar 15 08:29	Mar 22 18:44	Mar 30 20:08	Apr 06 23:37
	Apr 13 19:23	Apr 21 14:15	Apr 29 08:40	May 06 05:19
	May 13 07:20	May 21 08:29	May 28 18:42 n	Jun 04 10:38
	Jun 11 20:41 A	Jun 20 00:28	Jun 27 02:52 n	Jul 03 17:02
	Jul 11 11:16	Jul 19 13:59	Jul 26 09:58	Aug 02 01:51
	Aug 10 02:36	Aug 18 01:09	Aug 24 16:57	Aug 31 14:04
	Sep 08 18:09	Sep 16 10:20	Sep 23 00:54	Sep 30 06:01
	Oct 08 09:28	Oct 15 18:03	Oct 22 10:56	Oct 30 01:08
	Nov 07 00:14	Nov 14 01:07	Nov 20 23:50 n	Nov 28 22:06
	Dec 06 14:05 A	Dec 13 08:38	Dec 20 15:41	Dec 28 19:10

Year	New Moon	First Quarter	Full Moon	Last Quarter
2068	Jan 05 02:38	Jan 11 17:47	Jan 19 09:45	Jan 27 14:27
	Feb 03 13:44	Feb 10 05:20	Feb 18 04:38	Feb 26 06:25
	Mar 03 23:38	Mar 10 19:26	Mar 18 22:56	Mar 26 18:20
	Apr 02 08:51	Apr 09 11:33	Apr 17 15:29	Apr 25 02:30
	May 01 18:07	May 09 04:47	May 17 05:35 p	May 24 08:00
	May 31 04:03 H	Jun 07 22:20	Jun 15 17:00	Jun 22 12:25
	Jun 29 15:11	Jul 07 15:31	Jul 15 02:07	Jul 21 17:22
	Jul 29 03:55	Aug 06 07:38	Aug 13 09:51	Aug 20 00:16
	Aug 27 18:28	Sep 04 22:04	Sep 11 17:19	Sep 18 10:16
	Sep 26 10:48	Oct 04 10:23	Oct 11 01:39	Oct 18 00:00
	Oct 26 04:17	Nov 02 20:38	Nov 09 11:40 t	Nov 16 17:33
	Nov 24 21:42 P	Dec 02 05:21	Dec 08 23:42	Dec 16 14:11
	Dec 24 13:44	Dec 31 13:23		

All times are in Universal Time (UT1).
Eclipses at New Moon (solar eclipse) or Full Moon (lunar eclipse), are indicated by these symbols:

Solar Eclipses	Lunar Eclipses
T – Total	t – Total (Umbral)
A – Annular	p – Partial (Umbral)
H – Hybrid	n – Penumbral
P – Partial	

Phases of the Moon

Year 2069	New Moon	First Quarter	Full Moon	Last Quarter
			Jan 07 13:43	Jan 15 12:16
	Jan 23 03:36	Jan 29 21:39	Feb 06 05:29	Feb 14 09:27
	Feb 21 15:17	Feb 28 06:54	Mar 07 22:35	Mar 16 03:31
	Mar 23 01:13	Mar 29 17:34	Apr 06 16:13	Apr 14 17:21
	Apr 21 09:58 P	Apr 28 05:56	May 06 09:11 t	May 14 03:10
	May 20 18:06 P	May 27 20:09	Jun 05 00:19	Jun 12 09:56
	Jun 19 02:14	Jun 26 12:10	Jul 04 13:05	Jul 11 14:59
	Jul 18 11:13	Jul 26 05:30	Aug 02 23:44	Aug 09 19:41
	Aug 16 22:03	Aug 24 23:17	Sep 01 09:06	Sep 08 01:22
	Sep 15 11:35	Sep 23 16:23	Sep 30 18:09	Oct 07 09:20
	Oct 15 04:03 P	Oct 23 07:57	Oct 30 03:35 t	Nov 05 20:40
	Nov 13 22:38	Nov 21 21:31	Nov 28 13:46	Dec 05 12:03
	Dec 13 17:38	Dec 21 09:00	Dec 28 00:50	

Year 2070	New Moon	First Quarter	Full Moon	Last Quarter
				Jan 04 07:16
	Jan 12 11:22	Jan 19 18:31	Jan 26 12:59	Feb 03 04:46
	Feb 11 02:52	Feb 18 02:33	Feb 25 02:31	Mar 05 02:11
	Mar 12 15:52	Mar 19 09:53	Mar 26 17:31	Apr 03 21:23
	Apr 11 02:30 T	Apr 17 17:32	Apr 25 09:31 n	May 03 13:11
	May 10 11:08	May 17 02:30	May 25 01:37	Jun 02 01:26
	Jun 08 18:24	Jun 15 13:40	Jun 23 16:57	Jul 01 10:33
	Jul 08 01:14	Jul 15 03:26	Jul 23 07:02	Jul 30 17:17
	Aug 06 08:51	Aug 13 19:40	Aug 21 19:54	Aug 28 22:41
	Sep 04 18:29	Sep 12 13:44	Sep 20 07:47	Sep 27 04:02
	Oct 04 07:01 A	Oct 12 08:40	Oct 19 18:59 p	Oct 26 10:47
	Nov 02 22:42	Nov 11 03:20	Nov 18 05:40	Nov 24 20:20
	Dec 02 16:53	Dec 10 20:32	Dec 17 16:05	Dec 24 09:31

All times are in Universal Time (UT1).
Eclipses at New Moon (solar eclipse) or Full Moon (lunar eclipse), are indicated by these symbols:

Solar Eclipses	Lunar Eclipses
T – Total	t – Total (Umbral)
A – Annular	p – Partial (Umbral)
H – Hybrid	n – Penumbral
P – Partial	

Photo 3–2 The changing phases of the Moon. ©2010 F. Espenak

The phases of the Moon for the entire 21st Century can be found at the AstroPixels.com website:

www.astropixels.com/ephemeris/moon/phases2001gmt.html

Books by the Author

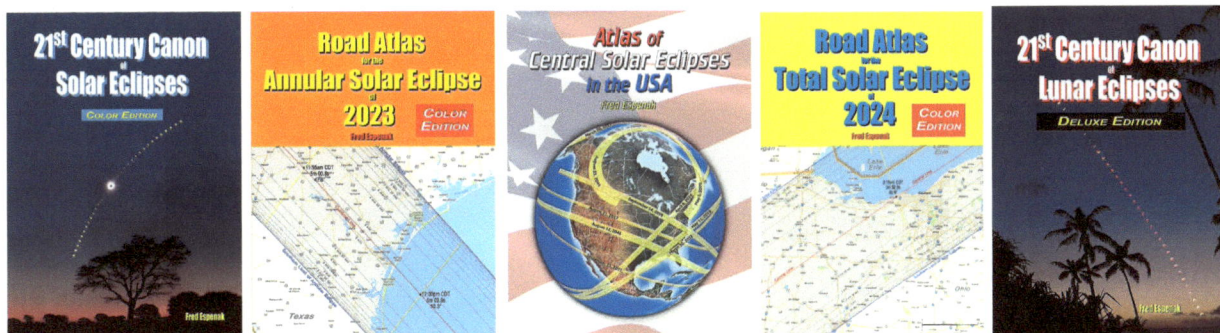

Below is a partial list of some of the books Fred Espenak has written through Astropixels Publishing:

21st Century Canon of Solar Eclipses

The complete guide to every solar eclipse occurring from 2001 tom 2100 (224 eclipses in all). It includes information and maps for all total, annular, hybrid, and partial eclipses. A special world atlas shows detailed full page maps of all central eclipse paths (total, annular and hybrid).

Road Atlas for the Annular Solar Eclipse of 2023

Detailed road maps of the entire eclipse path from the western USA, through Mexico, Central and South America. Information printed on the maps makes it easy to estimate the duration of annularity from any location in the eclipse path. This is the next annular solar eclipse visible from the USA.

Atlas of Central Solar Eclipses in the USA

When was the last total eclipse through the USA and when is the next? How often do they happen? What total eclipse tracks passed across the USA during the 17th, 18th, and 19th centuries, etc., and what states did they include? And how often is a total solar eclipse visible from each of the 50 states? The Atlas of Central Solar Eclipses in the USA answers all of these questions and more with hundreds of maps and tables.

Road Atlas for the Total Solar Eclipse of 2024

This book contains detailed road maps of the entire eclipse path from Mexico, through the USA and Canada. Information printed on the maps makes it easy to estimate the duration of totality from any location in the eclipse path. This is the next total solar eclipse visible from the USA.

21st Century Canon of Lunar Eclipses

The complete guide to every lunar eclipse occurring from 2001 tom 2100 (228 eclipses in all). It includes information and maps for all total, partial, and penumbral eclipses. The predictions use a new model for Earth's elliptical shadows.

For information on these books and more, visit *Astropixels Publishing*:

http://eclipsewise.com/pubs/index.html

www.ingramcontent.com/pod-product-compliance
Lightning Source LLC
Chambersburg PA
CBHW060817270326
41930CB00002B/73